Environmental Footprints and Eco-design of Products and Processes

Series Editor

Subramanian Senthilkannan Muthu, Chief Sustainability Officer
Green Story Inc.
Toronto, ON, Canada

Indexed by Scopus

This series aims to broadly cover all the aspects related to environmental assessment of products, development of environmental and ecological indicators and eco-design of various products and processes. Below are the areas fall under the aims and scope of this series, but not limited to: Environmental Life Cycle Assessment; Social Life Cycle Assessment; Organizational and Product Carbon Footprints; Ecological, Energy and Water Footprints; Life cycle costing; Environmental and sustainable indicators; Environmental impact assessment methods and tools; Eco-design (sustainable design) aspects and tools; Biodegradation studies; Recycling; Solid waste management; Environmental and social audits; Green Purchasing and tools; Product environmental footprints; Environmental management standards and regulations; Eco-labels; Green Claims and green washing; Assessment of sustainability aspects.

Subramanian Senthilkannan Muthu
Editor

Energy Footprint and Sustainability

Case Studies

Springer

Editor
Subramanian Senthilkannan Muthu
Green Story Inc, Canada
Islands District, Hong Kong

ISSN 2345-7651 ISSN 2345-766X (electronic)
Environmental Footprints and Eco-design of Products and Processes
ISBN 978-3-031-63056-9 ISBN 978-3-031-63057-6 (eBook)
https://doi.org/10.1007/978-3-031-63057-6

This Springer imprint is published by the registered company Springer Nature Switzerland AG
The registered company address is: Gewerbestrasse 11, 6330 Cham, Switzerland

This book is dedicated to:
The lotus feet of my beloved Lord
Pazhaniandavar
My beloved late Father
My beloved Mother
My beloved Wife Karpagam and Daughters
—Anu and Karthika
My beloved Brother
Last but not least
To everyone working in the energy sector to
make it "SUSTAINABLE"

Contents

Introduction to the Book—Energy Footprint and Sustainability: Case Studies 1
Subramanian Senthilkannan Muthu

A Review of Measurement of Energy Footprints Across Different Industrial Sectors 5
Pradeep Kautish, Mujahid Siddiqui, Nirzar Kulkarni, and Aaliyah Siddiqui

Promoting a Sustainable Energy Footprint: Brazilian Initiatives for Renewable Hydrogen 23
Victor Hugo Souza de Abreu and Andrea Souza Santos

The Urban Energy Footprint Model 37
Anthony M. Yezer

Exploring the Intersection of Human Psychology, Business Analysis, and Project Management to Reduce Computer Equipment's Environmental Impact 65
Richard Govada Joshua

Providing Decent, Affordable, and Sustainable Housing: Analysing Environmental Impacts of Family Houses Built with Conventional and Unconventional Building Materials 105
Aron Szennay, Zoltán Major, Róbert Horváth, Herta Czédli, and Cecília Szigeti

Index 127

Introduction to the Book—Energy Footprint and Sustainability: Case Studies

Subramanian Senthilkannan Muthu

Energy is one of the main elements under the umbrella of sustainability. Assessment of the energy and reduction of energy is an inevitable part of the sustainability activity in any industrial sector. Energy footprint quantifies the impact of human energy activities by tracking the total energy use of a product, process, organization, or even a service. As we are experiencing ever-increasing evidence of global warming caused by burning various fossil fuels, there is an overarching interest from various industrial players and any energy user about energy footprint or energy sustainability are shifting towards environmental dimensions of energy conservation, such as improving the energy efficiency and switching to renewable energy options.

This book deals with the concept of energy footprints. It begins with the concept of energy footprint and its implications on sustainability, followed by the details pertaining to the overview of models and methods for the measurement of energy footprints across different industrial sectors, and presents the applications and concepts of energy footprint measurement from a few industrial case studies.

This book is dedicated to focusing on the implications of energy footprint on sustainability, and this important aspect is showcased with five very informative chapters.

Chapter "A Review of Measurement of Energy Footprints Across Different Industrial Sectors" begins with the acknowledgment that various researchers and institutions have contributed to the development of metrics which measure the energy footprints of different industrial sectors. This study (Chapter "A Review of Measurement of Energy Footprints Across Different Industrial Sectors") aims to review the metrics in the current body of knowledge to showcase a structured analysis of various indicators which determine the energy footprints. Additionally, the study captures the complex set of indicators and metrics that are applied across

S. S. Muthu (✉)
Green Story Inc., Toronto, ON, Canada

S. S. Muthu (ed.), *Energy Footprint and Sustainability*, Environmental Footprints and Eco-design of Products and Processes, https://doi.org/10.1007/978-3-031-63057-6_1

different industries for measuring the energy footprint, which is one of the important aspects of this chapter. The authors of chapter "A Review of Measurement of Energy Footprints Across Different Industrial Sectors" have made an attempt to identify the leading models and methods for the measurement of energy footprints across different industrial sectors. In doing so, they aimed to help create an updated knowledge base for understanding the current development in the field of assessment of the energy footprint. This can also be of interest to scholars and professionals in different industrial sectors to identify the scope of development of these methods in improving the current models and methods. This review, by the authors of chapter "A Review of Measurement of Energy Footprints Across Different Industrial Sectors", attempts to identify the models used in the four large industrial sectors, namely, agriculture, construction, manufacturing, and textile. The main reason for selecting these are that these sectors are considered predominantly the largest contributors to the energy footprints. The authors conclude their review with some interesting points for each of the selected sectors.

Chapter "Promoting a Sustainable Energy Footprint: Brazilian Initiatives for Renewable Hydrogen" revolves around a compilation of governmental actions (programmes, partnerships, road maps, plans, resolutions, and decrees) related to the promotion of renewable hydrogen in Brazil. The crux of this study is to carry out a literature review to identify the main Brazilian government actions that seek to promote the use of renewable hydrogen in the country, boosting the development of national technologies, increasing the country's energy security—diversification of the Brazilian energy matrix—and creating new business opportunities and jobs in the hydrogen production chain. In addition, special emphasis has been given to the main barriers encountered in Brazil, as well as the main strategies needed to minimize them. The results show that efforts have been implemented since the 2000s until the establishment of the National Hydrogen Program (PNH2). During the literature review, several obstacles involving technological, financial, regulatory, infrastructure, and public awareness aspects were identified. To overcome these obstacles, a joint effort between the government, industry, and society is necessary through investment in research and development, implementation of effective public policies, stimulation of the adoption of more sustainable technologies, and promotion of awareness about the advantages of the use of renewable hydrogen. The authors conclude that given the need for further studies on the subject, it is recommended that research be carried out that takes a more in-depth look at the technical, economic, and environmental feasibility of using renewable hydrogen on a large scale in Brazil.

Chapter "The Urban Energy Footprint Model" is all about the urban energy footprint model. According to the authors, there is a long tradition extending since 1975 for modeling land use in cities using numerical urban simulation models. The techniques for solving these models have advanced considerably, and today, they can generate counterfactual estimates of the long-run effects of changes in many policies involving transportation, zoning, public infrastructure changes, etc. Several researchers modified these models to simulate the effects of transportation policies on energy consumption in commuting and housing, and the urban energy footprint

(UEF) model was born. Since then, the model has been adapted to consider the energy implications of self-driving cars, telework, planning regulations, ride hailing, and work from home among other applications discussed in detail in this chapter. The authors conclude that the UEF model is slightly more than a decade old. The approach followed has been applied to a variety of changes, ranging from high occupancy vehicle (HOV) lanes to telework to determine the effects on both urban form and carbon footprints. The authors have also listed some merits of the UEF models, such as the considering household utility effects.

Chapter "Exploring the Intersection of Human Psychology, Business Analysis, and Project Management to Reduce Computer Equipment's Environmental Impact" is dedicated to investigating the complex interaction between computer equipment's energy footprint and convergence of human psychology, business analysis, and project management approaches. This chapter aims to examine how organizational and individual decision-making affects energy use in the IT industry, drawing on concepts from behavioral psychology. By employing the application of a business analysis framework, this chapter identifies the primary variables that drive energy efficiency and suggests tactical approaches to reduce ecological footprint (EF). It addresses the challenges of adopting sustainable practices by using a project management approach, thereby highlighting the significance of cooperation and flexibility. According to the author of this chapter, to solve environmental concerns in the IT sector and promote sustainable practices, strategic solutions informed by interdisciplinary viewpoints must be applied.

Chapter "Providing Decent, Affordable, and Sustainable Housing: Analysing Environmental Impacts of Family Houses Built with Conventional and Unconventional Building Materials" deals with the analysis of the environmental impacts of family houses built with conventional and unconventional building materials. According to the authors of this chapter, the construction industry is considered an activity responsible for both significant CO_2 emissions and material flows. The authors state that providing affordable, decent, and sustainable habitats is, however, a fundamental human need. Therefore, this chapter aims to analyse the environmental impacts of construction and energetic upgrades of detached family dwellings typical in Hungary. For the analysis of environmental impacts, the concept of ecological footprint (EF) was used. The sample mentioned in this chapter consists of data from five detached family houses, representing the housing in Hungary: (1) a new two-storey family house with a gable roof, (2) a new family house made of timber, and (3) an energetic upgrade of a typical family dwelling built between 1950 and 1990, called "Hungarian cube". According to the results presented in this chapter, the construction of new detached family dwellings has a significantly higher ecological footprint than renovating an existing one. Using nonconventional construction materials (e.g. timber), however, can decrease the ecological footprint of construction, but useful lifetime should be considered more deeply in this case. The authors conclude that volatile energy prices have highlighted the importance of the energetics of housing stock. Newly built dwellings have outstanding insulation and, therefore, have nearly zero emission during operation, but they are not the means of affordable housing due to their construction costs and the construction itself having a significant environmental impact, which is not outweighed by the energy savings during the useful lifetime. Therefore, retrofitting and even extension of dwellings can better contribute to the Sustainable Development Goals (SDGs). The authors have also listed the limitations of this study.

I take this opportunity to thank all the contributors for their earnest efforts to bring out this book successfully. I am sure the readers of this book will find it very useful.

A Review of Measurement of Energy Footprints Across Different Industrial Sectors

Pradeep Kautish, Mujahid Siddiqui, Nirzar Kulkarni, and Aaliyah Siddiqui

1 Introduction

Energy footprint, which is often measured using inventory data under the life cycle approach, tracks the overall energy usage of the item under consideration, such as product manufacturing, processes, organizational outputs, household activities, or regional connotations. It does this by monitoring the influence of human energy activities. The definitions that currently exist can be broadly classified into two groups: mass driven and land driven, according to their relevance and usage patterns. Yet, there is no universally acceptable definition of energy footprint (Pan et al., 2024). Energy is considered one of the key components associated with the production and industrial activities and thus has a direct impact on the economic development. The human dependence on energy for almost all developmental activities is inevitable, and as economic activities continue to thrive, the demand for energy shall continue to rise (Payab et al., 2023). As the economic activities continue to rise, the ill effects of energy produced from the fossilized sources have begun to surface. One of the foremost adverse effects of energy consumption patterns is in the form of global warming, which has been directly attributed to the burning of fossil and non-fossil fuels (Siddiqui et al., 2023b). The usage of energy for different industrial sectors shall continue to grow, and this cycle of growth and its negative consequences for the environment shall continue to pose an even more serious threat to the existence of planet Earth. It is, therefore, felt that the adoption

P. Kautish
Institute of Management, Nirma University, Ahmedabad, Gujarat, India

M. Siddiqui (✉) · N. Kulkarni
Dr. Ambedkar Institute of Management Studies and Research, Nagpur, Maharashtra, India

A. Siddiqui
Symbiosis Centre for Management Studies, Nagpur, Maharashtra, India

S. S. Muthu (ed.), *Energy Footprint and Sustainability*, Environmental Footprints and Eco-design of Products and Processes, https://doi.org/10.1007/978-3-031-63057-6_2

of sustainable technologies shall be an effective method for decreasing the energy footprints generated by several industrial sectors (Shahbaz et al., 2023).

As a first step in the direction of shifting to cleaner energy, it is imperative to identify the industrial sectors which have a greater contribution toward the energy footprints. Once the identification and assessments are done, this can help in the initiation of measures that can assist in reducing the energy footprints (Karamouz et al., 2022; Muthu, 2020b). This emphasizes the need to develop models and tools for the assessment of the energy footprints across different industrial sectors of the economy. These frameworks shall help in the identification of the processes, products, industries, regions, etc., that are the dominant forces generating higher energy footprints. This can help either prevent or find substitutes for these sources and create a much cleaner industrial environment. There have been concerted efforts by researchers, scientists, and professionals from the industry to develop tools and methods which can help in the measurement of energy footprints for various industrial sectors (Hren et al., 2023; Siddiqui et al., 2022; Sowby & Dicataldo, 2022; Xie et al., 2020). Energy footprints are environmental footprints that concentrate on just one environmental issue, namely, energy usage, similar to the carbon footprints. Therefore, the evaluation of the energy use associated with a designated product, organization, or region within a given temporal and spatial boundary is known as an energy footprint. Most techniques calculate this using a life cycle perspective for the purpose of calculating the energy footprints (Al-Mulali & Ozturk, 2015; Fasogbon & Igboabuchukwu, 2024; Pandey & Agrawal, 2014).

This review is an attempt to identify the leading models and methods for the measurement of energy footprints across different industrial sectors. In doing so, we help in creating an updated knowledge base for understanding the current development in the field of assessment of energy footprint. This can also be of interest to scholars and professionals in different industrial sectors to identify the scope of development of these methods in order to improve the current models and methods. This review attempts to identify the models used in the four large industrial sectors, namely, agriculture, construction, manufacturing, and textile. This is because these sectors are considered predominantly the largest contributors to the energy footprints (Muthu, 2020b; Onat & Kucukvar, 2020; Siddiqui et al., 2023a; Sobrino et al., 2016). Moreover, upon perusal of the studies in the domain of energy footprints, much of the literature is from these industrial sectors, which was helpful in preparing this review. This chapter discusses various models and methods for the measurement of energy footprints across four different industrial sectors and presents a conclusion to summarize the review.

2 Agriculture Industry

Approximately more than one-third of the greenhouse gas (GHG) emissions pertaining to carbon footprints that are now causing climate modifications are caused by agriculture. The majority of these emissions are related to industrial agriculture-introduced practices. Transforming farming is considered an essential preliminary step in reducing the effects of climate change, and multiple governments and citizens are focused on achieving current sustainable development targets. The Food and Agriculture Organization (FAO) of the United Nations predicts that by 2050, the amount of cultivable land per capita will drop to just one-third of what it was in 1970. There are many reasons for this decline, but the primary one is thought to be the world's fastest-growing human population, which is projected to reach 9.7 billion people by 2050. Today's agricultural systems, which are already under pressure to dramatically boost their output in order to feed the world's expanding population, are further stressed by factors such as climate change, dwindling fisheries, degraded soil, and rising agribusiness prices (Engler & Krarti, 2021).

The study by Al-Mansour and Jejcic (2017) proposed a conceptual model for the purpose of evaluating the energy footprint in agriculture and agro-based products. Although their model was a much-elaborated tool for measuring the entire carbon footprint, the energy footprint comprised a vital component of their proposed model. The proposed model was christened an AgrFootprint model and had a defined set of processes which shall be included in the estimation of the energy footprints. The mean usage of fuel-related data for both primary and secondary soil ploughing, planting, fertilization, preservation of plant produces, harvesting, and transport is the basis for the energy usage statistics utilized in AgrFootprint. The usage of a multi-furrow reversible plough plow is considered for primary soil tillage. Before seeding, a rotating machine or a rotary harrow is utilized for secondary soil digging. The AgrFootprint model is composed of a chain of parallel and serial operations that are coupled in the identical sequence as agricultural production. The average fuel consumption statistics for primary and secondary soil tillage, planting, fertilizing, applying plant-preserving products, harvesting, climate control, further drying, and transportation within the production unit is what defines the model's input data. A significant quantity of information on agricultural methods, kinds of equipment, usage of energy, utilization of inorganic and organic fertilizers, emission variables, etc., is needed for the model.

In the context of measuring energy usage in the agricultural sector, the study by Platis et al. (2019) suggested that life cycle assessment (LCA), which is predominantly used for carbon footprint and water footprint, can also be deployed for the energy footprints in the agriculture farms. In the context of dairy farms, Pagani et al. (2016) measured the energy footprints by drawing a sample of 15 farms by performing a comparative study of farms from the USA and Italy. This study examined three farming systems where different constitutions of the animal feed are classified as grain based, forage based, and organic feed. The system boundary included all of the following: machinery and equipment; forage and food cultivated on farmland for milking cows and cattle; and forage and crops bought from outside sources. Indirect energy inputs included all of the following: fuels and electricity used at the farm level; and barns, feed stores, warehouses, and parlors. Two factors were left out of the analysis: (i) the energy utilized in bringing supplies and tools to the farm because it was impossible to track every logistical aspect of production input and (ii) peripheral inputs, like the energy required to construct manufacturing plants that made machinery or fertilizer or the banks and legal firms that handled loan approvals and contract arrangements. Since these two variables may account for as much as 10% of overall energy expenses, around 90% of all inputs are captured in the study's boundary. Since the study intended to consider the production of unprocessed milk, the 1 kg of energy-corrected milk available at the end of the production cycle was used as the unit of measured output. Additionally, direct and indirect energy consumption by the machinery and equipment included within the boundaries defined earlier was estimated based on the established standards of consumption. This was also followed by recording the farming practices in the sample and how these practices can be improved to ensure an improvement in energy usage and reduction of energy footprint.

Another significant contribution to the process of measurement of energy footprint in the agriculture domain was the modified proposition to the life cycle assessment (LCA) method. The study by Van Der Werf et al. (2020) has proposed the modification by highlighting the shortcomings of LCA and the introduction of additional measures for the assessment of energy footprints in the agriculture sector. Their proposition contends that the current LCA research and approaches frequently support more input-intensive farm processes while misrepresenting lower-impact agroecological systems, such as organic farming. This is caused in part by LCA's product-driven methodology, which emphasizes biomass production over other ecological services provided by farms, and in part by the fact that LCA seldom takes into account factors that agroecology seeks to enhance (soil richness, biodiversity condition, and effects of consumption of pesticides). Intensive agricultural systems are further favoured by the prevalent norm of restricting the analysis of indirect influences in LCA studies to indirect land use change, employing economic models that overlook determinants of social shifts and the influence of environmental policy interventions. The study by Van Der Werf et al. (2020) proposed that in addition to the existing model of LCA, some new parameters should be added, including the impact of degradation of soil, negative effects on biodiversity, and ill effects of pesticide usage.

Thus, it is clearly evident that the measurement of energy footprints in the agricultural segment is still in the stage where new models and tools are required to address the multitude of issues involved.

3 Construction Industry

One of the foremost methods of evaluation of the energy footprints in the construction business is the building life cycle (BLC) method, which considers two prominent aspects: embodied and operational energy (Dixit et al., 2012). Around 40% of the energy used worldwide is consumed each year by buildings, building materials, and building components during various stages of their life cycles, including manufacture, acquisition of building equipment, construction material, energy usage, and demolition. Both embodied and operating energies make up a building's total life cycle energy. While operational energy is needed to run the building's operations, such as air-conditioning, illumination, and other apparatus, embodied energy is the total amount of energy utilized during the building's formation, restoration, replacement, and destruction phases. Operational energy makes up a comparatively bigger share of a building's overall life cycle energy than embodied energy does. Previous studies have identified the relative contribution of embedded energy to total energy and have highlighted its significance, in light of the need to build environmentally friendly structures in the future (Plank, 2008). Several methodologies have been formed to appraise the environmental performance of structures, partly in reaction to the recognition of the methodological challenges and specifically connected to environmental consequences in the construction sector. Buildings are rated by the Green Building Rating Systems (GBRS) according to a range of evaluation criteria, such as the use of new resources potential reductions in energy, fuel, and water usage, and the quality of the interior system. Globally, GBRS has registered phenomenal growth, gained prominence, and taken a variety of forms through adaptations (Sartori et al., 2021). Embodied energy has the potential to be a useful tool for evaluating energy footprints because it is a credible measure of the generation of greenhouse gases. Nevertheless, an absence of accurate and uniform data as well as a paucity of common techniques beset the measuring methods in use today. Unreliable outcomes and data fluctuation can be attributed to multiple factors. A study of the literature conducted in recent research (Dixit, 2017) has shown that an embodied energy protocol should be created in order to facilitate the embodied energy analysis procedure. The Society for Environmental Toxicology and Chemistry (SETAC) published LCA guidelines that served as the basis for ISO LCA standards, as can be seen by looking through the standards that are currently in use. The LCA ISO standards have changed since then.

Another significant method used for the purpose of evaluation of the energy footprints was suggested in the form of Sustainability Development Index (SDI). At the 1992 Rio de Janeiro Environment and Development Conference, the United Nations wanted all nations to evolve their sustainability development indicators in order to support better decision-making at all levels (Barrera-Roldán & Saldívar-Valdés, 2002). Their study used the Figure of Merit to present the Sustainability Development Index by considering both socioeconomic characteristics of the human population and environmental factors specific to a location when establishing the SDI. The result was a long-term equilibrium between the natural order and habitation in a certain area. Furthermore, the outcomes helped decision-makers reach conclusions that supported sustainability. Later, a new ranking system was created by researchers just for structures taller than 20 stories, known as the Tall Building Projects Sustainability Indicator (TPSI) (Nguyen & Altan, 2012). Without accounting for the characteristics of materials used in construction, their method computed credits under the Building and Environment performance categories based on 119 basic variables. The newly proposed SDI makes it possible to carry out a variety of tasks, such as evaluating the environmental sustainability of a project's components and assisting in the efficient selection of building supplies, technologies, and design techniques that will minimize the energy impact of structures. Whenever that applies to the entire building system, SDI aids in grading a structure's sustainability status in relation to a benchmark building with the lowest possible energy usage. Buildings that are completed provide inputs for SDI evaluation. To calculate a building's sustainability, 11 main construction phase criteria are taken into account: formwork, doors, walls, windows with glazing, waterproofing, concrete, painting, reinforcement, plaster, structural steel, and earthwork. These fundamental metrics are also helpful in determining a building's energy footprints. The foremost advantage offered by the SDI method of estimating the energy footprints is that this can be done prior to the actual commencement of the construction works and can be helpful in reducing the energy footprints right at the conception stage of the project. It also makes it possible to lower the impact of emissions and embodied energy by utilizing alternate building materials, techniques, and technology. The procedure is, in theory, ubiquitous and can be used at different phases of any infrastructure or building project (Sabnis & Pranesh, 2017).

Sangireddy et al. (2019) presented a novel approach for developing effective, programmed sampling methods that, when trained on a regression-based modeling approach, will fit the data pertaining to the consumption of energy. Their findings indicate that when the sampled data is accomplished on an SVR model with a Gaussian RBF kernel, the data sampled by k-means clustering retains the energy distribution and yields the best outcomes in terms of accuracy. The study used histogram plots to confirm that the top distinctive feature points, extracted using k-means clustering, continue with the pattern of the energy consumption distribution. Several criteria were employed to assess the model's accuracy following the application of regression models for the forecasting of energy footprints. By contrasting the forecasts through a simulation, the models are verified. This model's benefit for calculating energy footprints is that it simulates energy use at a mere ~1% of the baseline data and produces results close to 95% accuracy. This study used the sample data for two different cities for the purpose of testing the model for measuring the energy footprints in the construction industry. This model will be particularly advantageous to buildings which are still at the design stage.

Having said that, we must recognize the enormous efforts that multiple national and international organizations have put in the past to create a range of guidelines, methods, and instruments for calculating the energy footprint of the construction, building, and infrastructure sectors. However, adoption of these technologies has been slow for a number of reasons, including the industry's inherent intricacy and character. Furthermore, the likelihood of further lowering footprints may be jeopardized because construction businesses want to only publish the emissions over which they have complete possession and knowledge. The building industries' efforts at implementing management solutions for reducing their energy footprint would be hampered by these and other such issues. Nevertheless, since 2007, the utilization of LCA research has grown dramatically, and since 2014, interest in the hybrid approach has grown (Fenner et al., 2018). However, this was contested in recent studies (Onat & Kucukvar, 2020) which assert that, if contrasted with the remaining two approaches, hybrid LCA and multi-regional IOA, the input-output technique of evaluation is the most widely used approach for estimating energy footprint in the construction industry. Still, the measurement models for evaluating the energy footprints in the construction industry continue to remain challenging due to the multiplicity of variations of construction projects (Denwigwe et al., 2019). Hence, the measurement models need a large amount of modifications based on the distinctiveness and scope and limits of construction projects (Dixit, 2017; Labaran et al., 2022).

4 Manufacturing Industry

A nation's ability to grow economically sustainably is greatly influenced by its manufacturing sector. Since the 2008 financial crisis, the manufacturing sector across the world has had a consistent growth in economic output. The industrial sectors' steady economic growth is accompanied by challenges related to the availability of natural assets and energy, as well as ecological strains, such as growing carbon, energy, material, and land footprints. Planning and policy formulation for sustainable development must take into account the manufacturing sectors' usage of renewable and nonrenewable energy. Industrial development boosts economic growth; however, the exact effect varies according to the growth level of the nation (Haraguchi et al., 2017). The degree to which new technologies are adopted, as well as the application of smart ways of manufacturing and the mitigation of their effects, will determine how quickly industrial growth occurs across the nations. These mitigation strategies will be essential to lessen the impact of economic development on environmental protection and climate both at the local and global levels (Ringel et al., 2016). Industrial development for developing nations entails a structural reformation by shifting from traditional industries like farming and fishing towards a contemporary manufacturing sector driven by innovation and technology (Szirmai, 2012). Such growth in the manufacturing sector produces necessary goods and services for consumers, fosters and encourages new technology, helps eliminate poverty in society, and generates employment (Rodrik, 2014). The primary source of industrial greenhouse gases is energy generation, either directly through the burning of petroleum or natural gas or through the usage of electricity. Due to the technology employed and the nature of production, lower-middle-income countries usually have the greatest emissions intensities; therefore, environmentally friendly industrialization pathways are especially pertinent to the growth trajectories of these nations (Mundaca & Markandya, 2016). The manufacturing industry has, therefore, emerged as one of the leading sources of energy and carbon footprints in both developing and developed economies of the world.

The manufacturing industries are also feeling the need to assess and measure the energy footprints of their industrial processes and production functions. The researchers in this domain have also been able to develop models and techniques for the purpose of assessment. For numerous components made of hard-to-cut metals, high-precision machining is becoming more and more crucial in modern manufacturing organizations, including aviation, healthcare, and automobiles. In response to these requirements, machine toolmakers have developed increasingly sophisticated tools and equipment to achieve wider adaptability and high accuracy at the cost of energy conservation. Regrettably, the world's vital energy supplies are still running low. Additionally, with many developing nations experiencing increasing financial growth, greenhouse gas emissions, which are largely tied to energy, are becoming a global concern.

Therefore, researchers have developed several models and measurement techniques which can help in estimating energy footprints across different manufacturing processes. One of the foremost in the case of other industries is the LCA model. As is typical of the LCA methodology, the manufacturing sector also has several challenges, with its usage for measuring the energy footprints. Meanwhile, a data surge caused by the growing usage of sensors and Internet of Things (IoT) technology across factories has helped generate several applications of cutting-edge algorithms used in deep learning for the deployment of big data from the manufacturing industry. One such model is proposed in the recent study by Serin et al. (2020) based on the deep multi-layer perceptron (DMLP) algorithm. This model uses the experimentation design technique for obtaining the data and training the DMLP model. This model was able to measure the energy footprints with ~90% accuracy and was used for the machine-level analysis of the energy footprints. In some of the earliest proposed models on the measurement of energy footprints, Savulescu and Kim (2008) suggested a technique for examining how much energy is used in the food industry and looking into the interactions between the water and energy systems. Their model was an overview of integrated design principles to provide a methodical framework for plant water and energy assessments that the manufacturing sector can use. Similarly, in the manufacturing of the marble industry, one unique approach was propounded by Gazi et al. (2012). They first prepared an extensive list of inventory used in the processing plants. This was followed by a comprehensive energy evaluation, with the goal of measuring the energy inputs for samples of production and giving the background information required to optimize the use of raw materials, earth resources, water, and energy based on the examination of unit operation scenarios. Standalone operations were then aggregated to measure the combined effects of the energy footprints. Vikhorev et al. (2013) suggested a structure for energy control and monitoring for the manufacturing industry. Their platform provides the requirements for energy data interchange, performance assessment, and energy usage display in addition to enabling an in-depth and real-time analysis of energy footprint data. It raises the understanding of how each component of the industrial facility uses energy and enables us to study the combined effect. Another novel method based on the energy key performance indicators (e-kpi) was propounded by May et al. (2015). They provided a technique for creating energy-related and production-tailored e-kpis that make it possible to interpret cause-and-effect correlations. To improve the productivity of the operation, the indicators examine a number of areas of the manufacturing process for the purpose of estimating the energy footprints. A process of application of mathematical models for the measurement of energy footprint was proposed by Giacone and Mancò (2012). They proposed a single-matrix equation that defines the connection between imported energies and energy drivers and is used to depict the entire energy system of a location. Each process's unique energy consumption makes up the matrix's elements. Energy consumption data is statistically analyzed through mathematical process modeling to determine the precise energy usage as a prime function of output.

Benedetti et al. (2017) presented techniques for regulating the energy-led efficiency of manufacturing facilities by assessing particular indicators. The different processing steps are characterized by the indices, which also pinpoint the highest-consuming manufacturing operations, enabling the computation of energy footprints of the manufacturing operations. A somewhat similar process of evaluation was also suggested by Cosgrove et al. (2017) and Wang et al. (2013) using the KPI methodology for distributed assessment and then aggregated results for measuring the energy footprints in the manufacturing sector. One more methodology suggested by Finnerty et al. (2017) aimed at developing an energy footprint measurement mechanism for multi-site production units. Their approach integrates unique key performance indicators and benchmarking for quantitative assessment of performance, along with energy management maturity models for qualitative description. The approach ultimately results in a multi-criteria decision-making methodology that supports a methodical, predictable, and expanding supportive decision-making structure. Recent times have witnessed a slow growth in modelling the approaches for the estimation of energy footprints. The study by Abbood et al. (2023) proposed a novel method using the multi-region input-output approach for the analysis of the energy footprints. This method is distinct in that it takes into account unpredictability related to international trade and develops the MRIO using the stochastic approach. Data analytics and statistical modelling techniques have been used to assess the top energy-consuming and carbon-emitting industries, including the manufacturing sector. However, the study by Usubiaga-Liaño et al. (2021) identified the problems associated with the MRIO due to double accounting issues while assessing the energy footprints.

5 Textile Industry

The inherent processes and techniques involved in the production, use, and recycling of textiles also contribute to a variety of emissions and consumption of natural resources generating energy footprints. Generally speaking, the word "textiles" refers to a wide range of product categories and processing phases, from the stages of fibre production or manufacturing to the final product processing phases, which include the stages of garment production. Basically, using textile products for body covering and protection is unavoidably important. The usage of textiles has significantly developed the industrial processes of manufacturing, technological advancements, and development of multifunctional applications, thus, increasing the normal consumption potential of textiles and textile products. The world's yearly fibre consumption has reached over 80 million tonnes, or 11.8 kg per person, as a result of expanding novel markets for textile products (Palamutcu, 2015). Similarly, the textile industry—a major processing and production sector in the world—is dealing with a growing number of serious environmental issues; the industry as a whole carbon-neutral environmental protection has a big influence on climate change (Zhang et al., 2020). To enhance the ecological value of goods, it is, therefore, essential to measure the carbon footprint of the textile sector throughout its entire life cycle and the reduction of carbon emissions of the textile industry. Many attempts have been made by researchers and practitioners in the past to propose frameworks and methods to assess the environmental impact of the textile industries and its various production systems (Jhanji, 2023; Muthu, 2020a).

The studies in the domain of textiles have proposed several methods and models which are employed for the purpose of determination of energy footprint. One brief study (Butnariu & Avasilcai, 2014) presented a case for estimation of energy footprint in the textile sector by capturing data from three sets of sources used in the textile sector, namely energy, resources, and waste. This study proposed that the inclusion of electricity has to be distributed across different sources as it is not directly sourced from any single natural resource. Similarly, renewable energy sources are to be excluded from the estimates as these sources have a high productivity factor compared with fossil-based sources. This study ignored several other parameters, and, hence, the proposed conceptual model cannot be used for a comprehensive appraisal of the energy footprints, but yet provided a model of measurement for some quick estimations in the textile manufacturing process. A more elaborated process using the life cycle assessment (LCA)-based approach for the textile industry was proposed by Jhanji (2023). A standardized framework termed life cycle assessment (LCA) is used to measure how products, activities, and services affect the natural environment. Using life cycle assessment (LCA) tools and techniques, products' ecological impact can be measured in terms of energy consumption, greenhouse gas emissions, carbon footprints, and water contaminants. Furthermore, LCA provides a comprehensive viewpoint and is essential in raising the sustainability ranking of textiles because it emphasizes the full life cycle of the product, from the acquisition of raw materials to their production, shipment, consumption, refurbishment, regular upkeep, disposition, reuse, and recycling. The definition of sustainable development goals and scope, inventory analysis, impact assessment, and interpretation phase are the four stages that make up the LCA framework. Whereas LCA as an approach is commonly used in the textile industry, another study by Vetőné Mózner (2013) used input-output models for measuring the carbon footprints. This study does a relative examination of four European nations with significant trading and carbon footprints. The researcher used input-output models as they offer an ideal structure for accounting for climate change. To reduce the environmentally devastating load of production activities, countries would be forced to reconsider their trade practices and shift from a too much production-driven approach to one based on consumption and allocation. The study's findings demonstrate the significance of differentiation between the two emission accounting strategies on a local and global scale. The consumption-based approach as a model for measuring the carbon footprint has been effectively deployed by some of the previous studies (Adnan et al., 2018; Heinonen et al., 2020) in other contexts and has been proven to be an effective measurement tool.

Another important work by Muthu (2020b) has outlined the critical aspects of energy footprint measurement in the textile industrial sector. The supply chain of the textile manufacturing and process is highly dependent on fuel and power. Fuels are mostly used in the water-driven processing of textiles, whereas electricity is mostly required to operate various machines. There are, therefore, a multitude of challenges in modelling a single-measurement method for all the processes in the textile industry. The reasons for such challenges emerge from the fact that there are different types of fibres, different production methods, and different machinery depending on the output to be produced. For instance, the energy consumed for spinning, weaving, and wet processes hugely varies from each, and, hence, the energy consumption also varies widely. Similarly, different textile products also demand different intensities of energy consumption. This also creates challenges for modelling the mechanism for energy footprint measurement in the textile industry. Still, there are certain principles based on which the energy and fuel consumption can be estimated and the energy footprint can be measured in the textile industry. This can be done by tracking the direct and indirect energy requirements of a specific procedure. A combined demand for energy can be evaluated in the total supply chain or a modular approach for different operations can be adopted for evaluating the energy footprint. Power consumption and use can be evaluated for a given procedure, different processes in a supply chain, or for an aggregate product. The energy use can be measured in megajoules (MJ) or any other pertinent unit for a certain process of manufacturing, and suitable units of measurement can be employed. The life cycle assessment (LCA) approach, which tracks the amount of energy consumed in the whole product life cycle from cradle to grave phases, can be used to calculate a product's energy footprint or energy consumption. This approach takes into consideration the total energy needed to create a product, from the collection of raw materials to the dumping phase, including the energy from feedstock. This method takes into consideration the energy used during the production of the product and any related components as well as incidental operations like the extraction of raw materials and the production of power. It also considers how energy is used in various stages of the life cycle.

Another earlier work by Muthu (2014) perhaps did not directly mention the term "energy footprints" in their article, but obliquely suggested models which could measure the environmental influence of the textile industry. This article briefly proposed three different techniques of measurement of the environmental impact of the textile sector. The first technique relates to the life cycle assessment (LCA) which intends to measure the energy footprints right from the stage of extraction of raw materials, to the manufacture, consumption, transportation, and disposal. LCA is used to evaluate the effects of products from cradle to grave. To assess various potential environmental consequences, both positive and negative, the LCA technique focuses on gathering and analysing quantitative data on the numerous inputs and outputs of pertinent raw and semi-finished materials, energy flows, and waste movements. The second technique for the measurement of energy footprints was termed "material flow analysis" (Haberl et al., 2004). The term "material flow analysis (MFA)" is an analytical technique for methodically assessing material flows and stockpiles under precise conditions of production and supply. As a significant tool, it finds use in environmental engineering, waste management, resources management, and information engineering, among other fields. It can be used for a broad range of materials, chemicals, processes, and products in a variety of fields (Brunner & Rechberger, 2016). The third process that can be effectively deployed for the estimation of energy footprints in the textile sector was suggested to be the material intensity per service (MISP) unit. In the 1990s, the Wuppertal Institute in Germany established the concept of MIPS. MIPS, which measures the volume of resources utilized to provide a given good or service in kilogrammes per unit of service, aims to provide that information. The five categories into which MIPS divides the material sources are abiotic, biotic raw materials, earth movement used in silviculture and agriculture, water, and air (Ritthoff et al., 2002). A recent study (Imran et al., 2023) which measured the energy footprints of the textile sector in Pakistan adopted a survey-based method in combination with the data extracted from secondary sources. They gathered information on several steps taken in the manufacturing of industrial textiles, such as spinning the thread, weaving the yarn, knitting the cloth, processing, and clothing. The model created in this study offers a thorough simulation of the essential manufacturing processes frequently applied in production facilities ranging from small to large. The data from the research and survey was obtained in order to compute from specification data sheets or obtain readings from production equipment. To ensure the accuracy of the data, the total of each subprocess was compared to statistical information on the overall use of materials and energy by firms. In conclusion, as presented above, there are multiple techniques available for the purpose of measurement of energy footprints in the textile industries.

6 Conclusion

The current chapter aimed to identify the models and methods for the measurement of energy footprints for different industrial sectors. The aim was to perform an overview of the existing research and identify the most common methods. This was done with the aim of understanding the trends in different industry and in the process also understanding the direction in which researchers are working for further developing these models and methods. Researchers have suggested one uncommon model for measuring the energy footprints in the agriculture industry. The fuel data for both primary and secondary soil tillage, planting, fertilization, preservation of plant products, harvesting, and transport are the basis for the energy usage statistics utilized in the model titled AgrFootprint. Although this model takes a comprehensive view and provides detailed analysis, yet the collection of data for analysis remains a shortcoming. The other model intended to measure the energy footprints of agro farms also takes a comprehensive view by including direct and indirect sources of energy consumption in a farm environment. Although some researchers have used the life cycle assessment model, yet they have proposed a modified LCA model as adopted in the agriculture industry for improving the performance of the models.

The construction industry has suggested the approach of embodied energy analysis for capturing the true energy footprints. The Figure of Merit to develop a Sustainability Development Index (SDI) by considering both the socioeconomic characteristics of the human population and environmental factors specific to a location was given due consideration when establishing the SDI. Researchers working in the construction industry models also proposed a mathematical model for the estimation of energy footprints through a collection of sampled data and then used the sampled data for training an SVR model with a Gaussian RBF kernel for improving the predictability potential of energy footprint estimation models.

The manufacturing industry has seen a wider usage of big data in its energy footprint estimation models. In this industry, the usage of models that operate on deep learning by the deployment of big data from the manufacturing industry has been proposed by researchers. Models based on the deep multi-layer perceptron (DMLP) algorithm as well as stochastic models have also been proposed for the manufacturing industry. Other trends such as the KPI-based models and development of indices for preparing the prominent processes and products generating the largest energy footprints have been suggested for the manufacturing industry.

The textile industry has a wider variety of models for the measurement of energy footprints. The suggested model aimed at capturing data from three sets of sources used in the textile industrial usages: energy, resources, and waste. A distributed capture of energy that will help in making a more comprehensive evaluation of the energy footprint was suggested in a model. Multiplicity of products and raw material and variation in the process of production and operations of different textile products have been highlighted by researchers as the main challenge for energy footprint measurements in the textile sector. The usage of the LCA model adapted for the textile industry was also proposed in the research. Material flow analysis and material intensity per service are two other innovative models which have been proposed in the literature on the measurement of energy footprints in the textile industry. A simulation-based model for the measurement was also proposed, which offers a thorough simulation of the essential manufacturing processes frequently applied in production facilities, ranging from small to large processes in the textile industry.

References

Abbood, K., Egilmez, G., & Meszaros, F. (2023). Multi-region input-output-based carbon and energy footprint analysis of U.S. manufacturing. *Periodica Polytechnica Social and Management Sciences, 31*(2), Article 2. https://doi.org/10.3311/PPso.19554

Adnan, M. N., Safeer, R., & Rashid, A. (2018). Consumption based approach of carbon footprint analysis in urban slum and non-slum areas of Rawalpindi. *Habitat International, 73*, 16–24. https://doi.org/10.1016/j.habitatint.2017.12.012

Al-Mansour, F., & Jejcic, V. (2017). A model calculation of the carbon footprint of agricultural products: The case of Slovenia. *Energy, 136*, 7–15. https://doi.org/10.1016/j.energy.2016.10.099

Al-Mulali, U., & Ozturk, I. (2015). The effect of energy consumption, urbanization, trade openness, industrial output, and the political stability on the environmental degradation in the MENA (Middle East and North African) region. *Energy, 84*, 382–389. https://doi.org/10.1016/j.energy.2015.03.004

Barrera-Roldán, A., & Saldívar-Valdés, A. (2002). Proposal and application of a sustainable development index. *Ecological Indicators, 2*(3), 251–256. https://doi.org/10.1016/S1470-160X(02)00058-4

Benedetti, M., Cesarotti, V., & Introna, V. (2017). From energy targets setting to energy-aware operations control and back: An advanced methodology for energy efficient manufacturing. *Journal of Cleaner Production, 167*, 1518–1533. https://doi.org/10.1016/j.jclepro.2016.09.213

Brunner, P. H., & Rechberger, H. (2016). *Handbook of material flow analysis: For environmental, resource, and waste engineers, second edition*. CRC Press.

Butnariu, A., & Avasilcai, S. (2014). Research on the possibility to apply ecological footprint as environmental performance indicator for the textile industry. *Procedia – Social and Behavioral Sciences, 124*, 344–350. https://doi.org/10.1016/j.sbspro.2014.02.495

Cosgrove, J., Doyle, F., Littlewood, J., & Wilgeroth, P. (2017). A methodology for electricity monitoring and targeting (M&T) in an Irish precision engineering SME. *International Journal of Sustainable Engineering, 10*(4–5), 233–240. https://doi.org/10.1080/19397038.2017.1317877

Denwigwe, I. H., Babatunde, O. M., Babatunde, D. E., Akintunde, T. J., & Akinbulire, T. O. (2019). A technical review on methods and tools for evaluation of energy footprints, impact on buildings and environment. In S. S. Muthu (Ed.), *Energy footprints of the bio-refinery, hotel, and building sectors* (pp. 47–81). Springer. https://doi.org/10.1007/978-981-13-2466-6_2

Dixit, M. K. (2017). Life cycle embodied energy analysis of residential buildings: A review of literature to investigate embodied energy parameters. *Renewable and Sustainable Energy Reviews, 79*, 390–413. https://doi.org/10.1016/j.rser.2017.05.051

Dixit, M. K., Fernández-Solís, J. L., Lavy, S., & Culp, C. H. (2012). Need for an embodied energy measurement protocol for buildings: A review paper. *Renewable and Sustainable Energy Reviews, 16*(6), 3730–3743. https://doi.org/10.1016/j.rser.2012.03.021

Engler, N., & Krarti, M. (2021). Review of energy efficiency in controlled environment agriculture. *Renewable and Sustainable Energy Reviews, 141*, 110786. https://doi.org/10.1016/j.rser.2021.110786

Fasogbon, S., & Igboabuchukwu, C. (2024). Real-time carbon footprint assessment based on energy consumption: A comprehensive review for future research prospects. *Renewable and Sustainable Energy Reviews, 192*, 114225. https://doi.org/10.1016/j.rser.2023.114225

Fenner, A. E., Kibert, C. J., Woo, J., Morque, S., Razkenari, M., Hakim, H., & Lu, X. (2018). The carbon footprint of buildings: A review of methodologies and applications. *Renewable and Sustainable Energy Reviews, 94*, 1142–1152. https://doi.org/10.1016/j.rser.2018.07.012

Finnerty, N., Sterling, R., Coakley, D., Contreras, S., Coffey, R., & Keane, M. M. (2017). Development of a global energy management system for non-energy intensive multi-site industrial organisations: A methodology. *Energy, 136*, 16–31. https://doi.org/10.1016/j.energy.2016.10.049

Gazi, A., Skevis, G., & Founti, M. A. (2012). Energy efficiency and environmental assessment of a typical marble quarry and processing plant. *Journal of Cleaner Production, 32*, 10–21. https://doi.org/10.1016/j.jclepro.2012.03.007

Giacone, E., & Mancò, S. (2012). Energy efficiency measurement in industrial processes. *Energy, 38*(1), 331–345. https://doi.org/10.1016/j.energy.2011.11.054

Haberl, H., Fischer-Kowalski, M., Krausmann, F., Weisz, H., & Winiwarter, V. (2004). Progress towards sustainability? What the conceptual framework of material and energy flow accounting (MEFA) can offer. *Land Use Policy, 21*(3), 199–213. https://doi.org/10.1016/j.landusepol.2003.10.013

Haraguchi, N., Cheng, C. F. C., & Smeets, E. (2017). The importance of manufacturing in economic development: Has this changed? *World Development, 93*, 293–315. https://doi.org/10.1016/j.worlddev.2016.12.013

Heinonen, J., Ottelin, J., Ala-Mantila, S., Wiedmann, T., Clarke, J., & Junnila, S. (2020). Spatial consumption-based carbon footprint assessments—A review of recent developments in the field. *Journal of Cleaner Production, 256*, 120335. https://doi.org/10.1016/j.jclepro.2020.120335

Hren, R., Vujanović, A., Van Fan, Y., Klemeš, J. J., Krajnc, D., & Čuček, L. (2023). Hydrogen production, storage and transport for renewable energy and chemicals: An environmental footprint assessment. *Renewable and Sustainable Energy Reviews, 173*, 113113. https://doi.org/10.1016/j.rser.2022.113113

Imran, S., Mujtaba, M. A., Zafar, M. M., Hussain, A., Mehmood, A., Farwa, U. E., Korakianitis, T., Kalam, M. A., Fayaz, H., & Saleel, C. A. (2023). Assessing the potential of GHG emissions for the textile sector: A baseline study. *Heliyon, 9*(11), e22404. https://doi.org/10.1016/j.heliyon.2023.e22404

Jhanji, Y. (2023). Life cycle analysis of textiles and associated carbon emissions. In *Sustainable fibres for fashion and textile manufacturing* (pp. 403–431). Elsevier. https://doi.org/10.1016/B978-0-12-824052-6.00003-2

Karamouz, M., Zare, M., & Ebrahimi, E. (2022). *System dynamics based carbon footprint assessment of industrial water and energy use* [Preprint]. In Review. https://doi.org/10.21203/rs.3.rs-1770740/v1

Labaran, Y. H., Mathur, V. S., Muhammad, S. U., & Musa, A. A. (2022). Carbon footprint management: A review of construction industry. *Cleaner Engineering and Technology, 9*, 100531. https://doi.org/10.1016/j.clet.2022.100531

May, G., Barletta, I., Stahl, B., & Taisch, M. (2015). Energy management in production: A novel method to develop key performance indicators for improving energy efficiency. *Applied Energy, 149*, 46–61. https://doi.org/10.1016/j.apenergy.2015.03.065

Mundaca, L., & Markandya, A. (2016). Assessing regional progress towards a 'Green Energy Economy'. *Applied Energy, 179*, 1372–1394. https://doi.org/10.1016/j.apenergy.2015.10.098

Muthu, S. S. (2014). Ways of measuring the environmental impact of textile processing: An overview. In *Assessing the environmental impact of textiles and the clothing supply chain* (pp. 32–56). Elsevier. https://doi.org/10.1533/9781782421122.32

Muthu, S. S. (2020a). *Assessing the environmental impact of textiles and the clothing supply chain* (2nd ed.). Woodhead Publishing Elsevier.

Muthu, S. S. (2020b). Calculating the water and energy footprints of textile products. In *Assessing the environmental impact of textiles and the clothing supply chain* (pp. 77–93). Elsevier. https://doi.org/10.1016/B978-0-12-819783-7.00004-1

Nguyen, B. K., & Altan, H. (2012). Tall-building projects sustainability indicator (TPSI): A new design and environmental assessment tool for tall buildings. *Buildings, 2*(2), 43–62. https://doi.org/10.3390/buildings2020043

Onat, N. C., & Kucukvar, M. (2020). Carbon footprint of construction industry: A global review and supply chain analysis. *Renewable and Sustainable Energy Reviews, 124*, 109783. https://doi.org/10.1016/j.rser.2020.109783

Pagani, M., Vittuari, M., Johnson, T. G., & De Menna, F. (2016). An assessment of the energy footprint of dairy farms in Missouri and Emilia-Romagna. *Agricultural Systems, 145*, 116–126. https://doi.org/10.1016/j.agsy.2016.03.009

Palamutcu, S. (2015). Energy footprints in the textile industry. In *Handbook of life cycle assessment (LCA) of textiles and clothing* (pp. 31–61). Elsevier. https://doi.org/10.1016/B978-0-08-100169-1.00002-2

Pan, W., Hu, C., Huang, G., Dai, W., & Pan, W. (2024). Energy footprint: Concept, application and modeling. *Ecological Indicators, 158*, 111459. https://doi.org/10.1016/j.ecolind.2023.111459

Pandey, D., & Agrawal, M. (2014). Carbon footprint estimation in the agriculture sector. In S. S. Muthu (Ed.), *Assessment of carbon footprint in different industrial sectors, Volume 1* (pp. 25–47). Springer. https://doi.org/10.1007/978-981-4560-41-2_2

Payab, A. H., Kautish, P., Sharma, R., Siddiqui, A., Mehta, A., & Siddiqui, M. (2023). Does human capital complement sustainable development goals? Evidence from leading carbon emitter countries. *Utilities Policy, 81*, 101509. https://doi.org/10.1016/j.jup.2023.101509

Plank, R. (2008). The principles of sustainable construction. *The IES Journal Part A: Civil & Structural Engineering, 1*(4), 301–307. https://doi.org/10.1080/19373260802404482

Platis, D., Anagnostopoulos, C., Tsaboula, A., Menexes, G., Kalburtji, K., & Mamolos, A. (2019). Energy analysis, and carbon and water footprint for environmentally friendly farming practices in agroecosystems and agroforestry. *Sustainability, 11*(6), 1664. https://doi.org/10.3390/su11061664

Ringel, M., Schlomann, B., Krail, M., & Rohde, C. (2016). Towards a green economy in Germany? The role of energy efficiency policies. *Applied Energy, 179*, 1293–1303. https://doi.org/10.1016/j.apenergy.2016.03.063

Ritthoff, M., Rohn, H., & Liedtke, C. (2002). *Calculating MIPS: Resource productivity of products and services*. Wuppertal Institute for Climate, Environment and Energy. https://epub.wupper-inst.org/frontdoor/index/index/docId/1577

Rodrik, D. (2014). The past, present, and future of economic growth. *Challenge, 57*(3), 5–39. https://doi.org/10.2753/0577-5132570301

Sabnis, A., & Pranesh, M. R. (2017). Sustainability evaluation of buildings in pre-use phase using figure of merit as a new tool. *Energy and Buildings, 145*, 121–129. https://doi.org/10.1016/j.enbuild.2017.02.063

Sangireddy, S. A. R., Bhatia, A., & Garg, V. (2019). Development of a surrogate model by extracting top characteristic feature vectors for building energy prediction. *Journal of Building Engineering, 23*, 38–52. https://doi.org/10.1016/j.jobe.2018.12.018

Sartori, T., Drogemuller, R., Omrani, S., & Lamari, F. (2021). A schematic framework for life cycle assessment (LCA) and green building rating system (GBRS). *Journal of Building Engineering, 38*, 102180. https://doi.org/10.1016/j.jobe.2021.102180

Savulescu, L., & Kim, J.-K. (2008). 10—Novel methods for combined energy and water minimisation in the food industry. In J. Klemeš, R. Smith, & J.-K. Kim (Eds.), *Handbook of water and energy management in food processing* (pp. 304–331). Woodhead Publishing. https://doi.org/10.1533/9781845694678.2.304

Serin, G., Sener, B., Gudelek, M. U., Ozbayoglu, A. M., & Unver, H. O. (2020). Deep multi-layer perceptron based prediction of energy efficiency and surface quality for milling in the era of sustainability and big data. *Procedia Manufacturing, 51*, 1166–1177. https://doi.org/10.1016/j.promfg.2020.10.164

Shahbaz, M., Siddiqui, A., Siddiqui, M., Jiao, Z., & Kautish, P. (2023). Exploring the growth of sustainable energy Technologies: A review. *Sustainable Energy Technologies and Assessments, 57*, 103157. https://doi.org/10.1016/j.seta.2023.103157

Siddiqui, A., Siddiqui, M., & Kulkarni, N. (2022). Artificial Intelligence in water conservation: A meta-analysis study. *Prabandhan: Indian Journal of Management, 15*(3), 24–41. https://doi.org/10.17010/pijom/2022/v15i3/160407

Siddiqui, A., Altekar, S., Kautish, P., Fulzele, S., Kulkarni, N., Siddiqui, M., & Bashir, M. F. (2023a). Review of measurement of sustainable development goals: A comprehensive bibliometric and visualized analysis. *Environmental Science and Pollution Research, 30*(40), 91761–91779. https://doi.org/10.1007/s11356-023-28887-x

Siddiqui, A., Siddiqui, M., & Kautish, P. (2023b). Green finance and low carbon technology innovation: Current research and future outlook. In *Recent developments in green finance, green growth and carbon neutrality* (pp. 259–280). Elsevier. https://doi.org/10.1016/B978-0-443-15936-7.00015-3

Sobrino, N., Monzon, A., & Hernandez, S. (2016). Reduced carbon and energy footprint in highway operations: The highway energy assessment (HERA) methodology. *Networks and Spatial Economics, 16*(1), 395–414. https://doi.org/10.1007/s11067-014-9225-y

Sowby, R. B., & Dicataldo, E. (2022). The energy footprint of U.S. irrigation: A first estimate from open data. *Energy Nexus, 6*, 100066. https://doi.org/10.1016/j.nexus.2022.100066

Szirmai, A. (2012). Industrialisation as an engine of growth in developing countries, 1950–2005. *Structural Change and Economic Dynamics, 23*(4), 406–420. https://doi.org/10.1016/j.strueco.2011.01.005

Usubiaga-Liaño, A., Arto, I., & Acosta-Fernández, J. (2021). Double accounting in energy footprint and related assessments: How common is it and what are the consequences? *Energy, 222*, 119891. https://doi.org/10.1016/j.energy.2021.119891

Van Der Werf, H. M. G., Knudsen, M. T., & Cederberg, C. (2020). Towards better representation of organic agriculture in life cycle assessment. *Nature Sustainability, 3*(6), 419–425. https://doi.org/10.1038/s41893-020-0489-6

Vetőné Mózner, Z. (2013). A consumption-based approach to carbon emission accounting – Sectoral differences and environmental benefits. *Journal of Cleaner Production, 42*, 83–95. https://doi.org/10.1016/j.jclepro.2012.10.014

Vikhorev, K., Greenough, R., & Brown, N. (2013). An advanced energy management framework to promote energy awareness. *Journal of Cleaner Production, 43*, 103–112. https://doi.org/10.1016/j.jclepro.2012.12.012

Wang, Q., Liu, F., & Li, C. (2013). An integrated method for assessing the energy efficiency of machining workshop. *Journal of Cleaner Production, 52*, 122–133. https://doi.org/10.1016/j.jclepro.2013.03.020

Xie, J., Fu, J., Liu, S., & Hwang, W. (2020). Assessments of carbon footprint and energy analysis of three wind farms. *Journal of Cleaner Production, 254*, 120159. https://doi.org/10.1016/j.jclepro.2020.120159

Zhang, J., Qian, X., & Feng, J. (2020). Review of carbon footprint assessment in textile industry. *Ecofeminism and Climate Change, 1*(1), 51–56. https://doi.org/10.1108/EFCC-03-2020-0006

Promoting a Sustainable Energy Footprint: Brazilian Initiatives for Renewable Hydrogen

Victor Hugo Souza de Abreu and Andrea Souza Santos

1 Introduction

The large-scale deployment of renewable energy sources has been driven by the proposal of carbon-neutral targets in several countries, as well as intensified global action on climate change and the acceleration of the green economy in the post-pandemic period (Santos et al., 2021; Liu et al., 2022; Zhang et al., 2022; de Abreu et al., 2024). However, one of the biggest challenges that has prevented the widespread adoption of renewable energy sources, such as solar photovoltaic and wind power, is their intermittent nature, i.e., they are not constant and predictable in their energy production (Gulotta et al., 2022). Therefore, the construction of a hydrogen supply system has been increasingly accepted as a global consensus, since hydrogen has a higher heating value and lower carbon emissions than fossil fuels, according to studies such as Santos et al. (2021), Liu et al. (2022), Weidner et al. (2023), Capurso et al. (2022), Oliveira (2022) and De Abreu et al. (2023).

Currently, hydrogen production is mainly divided into three categories: (i) from renewable sources such as renewable biomass and biofuels, solar photovoltaics, wind energy and hydroelectricity (known as green hydrogen); (ii) from fossil sources with carbon capture and storage (CCS)/carbon capture, utilization and storage (CCUS) (known as blue hydrogen); and (iii) from fossil sources (known as gray hydrogen). To promote the development of the clean hydrogen market, renewable hydrogen standards, the first category, have been discussed at a global level (Bhandari et al., 2014; Liu et al., 2022). This energy source can be used in various

V. H. S. de Abreu (✉) · A. S. Santos
Federal University of Rio de Janeiro (Universidade Federal do Rio de Janeiro—UFRJ), Transport Engineering Programme (Programa de Engenharia de Transportes—PET), COPPE-UFRJ, Rio de Janeiro, Brazil
e-mail: victor@pet.coppe.ufrj.br; andrea.santos@pet.coppe.ufrj.br

S. S. Muthu (ed.), *Energy Footprint and Sustainability*, Environmental Footprints and Eco-design of Products and Processes, https://doi.org/10.1007/978-3-031-63057-6_3

sectors, such as transportation, industry and power generation, replacing fossil fuels and reducing greenhouse gas (GHG) emissions (Santos et al., 2021; Proença et al., 2023).

Therefore, the purpose of this study is to carry out a literature review to identify the main Brazilian government actions that seek to promote the use of renewable hydrogen in the country, boosting the development of national technologies, increasing the country's energy security—diversification of the Brazilian energy matrix—and creating new business opportunities and jobs in the hydrogen production chain. In addition, special emphasis is placed on the barriers that still exist, as well as some of the main ways of minimizing them. It is worth noting that this study is aligned with the Sustainable Development Goals (SDGs) of the 2030 Agenda for Sustainable Development, more specifically the SDGs listed in Fig. 1.

In addition to this introductory section, this chapter is structured as follows: Section 2 presents the methodology used in the research; Sect. 3 points out the main initiatives to implement renewable hydrogen on a large scale in Brazil, with emphasis on their objectives; Sect. 4 discusses the main barriers preventing the expansion of the use of renewable hydrogen in the country, as well as the possible solutions to minimize them; and finally, Sect. 5 presents the final considerations, where we reflect on the results obtained and point out the possible ways forward for the use of renewable hydrogen in Brazil, as well as suggesting new studies to delve deeper into the subject.

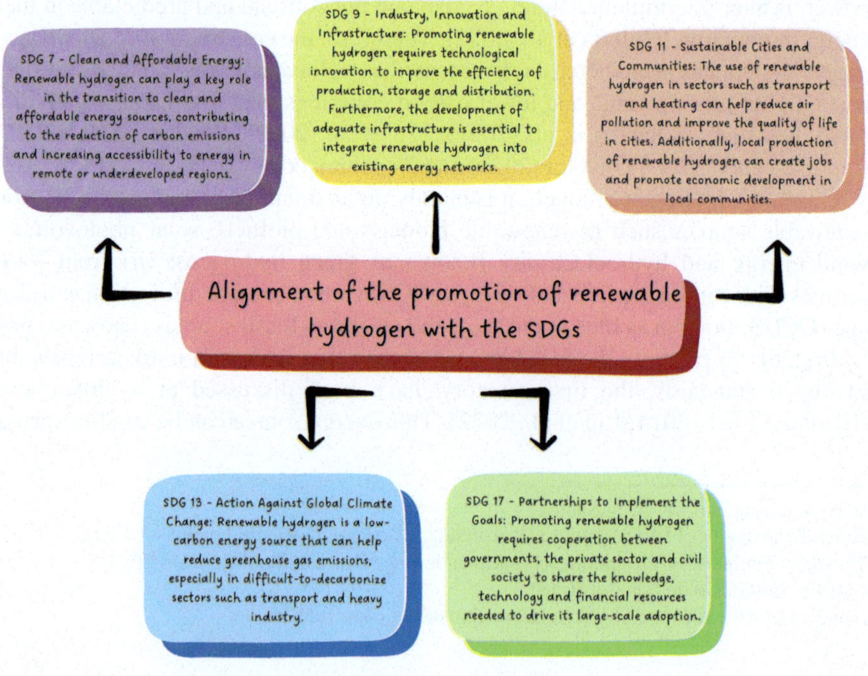

Fig. 1 Aligning the promotion of renewable hydrogen with the SDGs

2 Methodology

The methodology adopted in this study (as presented in Fig. 2) comprises a documentary review of the main policies and programs aimed at promoting renewable hydrogen in Brazil, as well as an analysis of the main difficulties faced in its implementation, which include the high cost of large-scale production, lack of infrastructure for transportation, storage and distribution, absence of specific regulations and adequate government incentives, scarcity of specialized technical knowledge and lack of collaboration between sectors and stakeholders, and ways of resolving them, or at least minimizing them.

3 Main Brazilian Initiatives

Brazil's geographical advantage positions it as a promising contender in the global hydrogen market. Leveraging its abundant renewable energy resources, the nation stands poised to not only meet domestic demand but also emerge as a significant player in the international arena. As underscored by Da Silva César et al. (2019), Brazil's robust renewable energy matrix confers a distinct competitive edge,

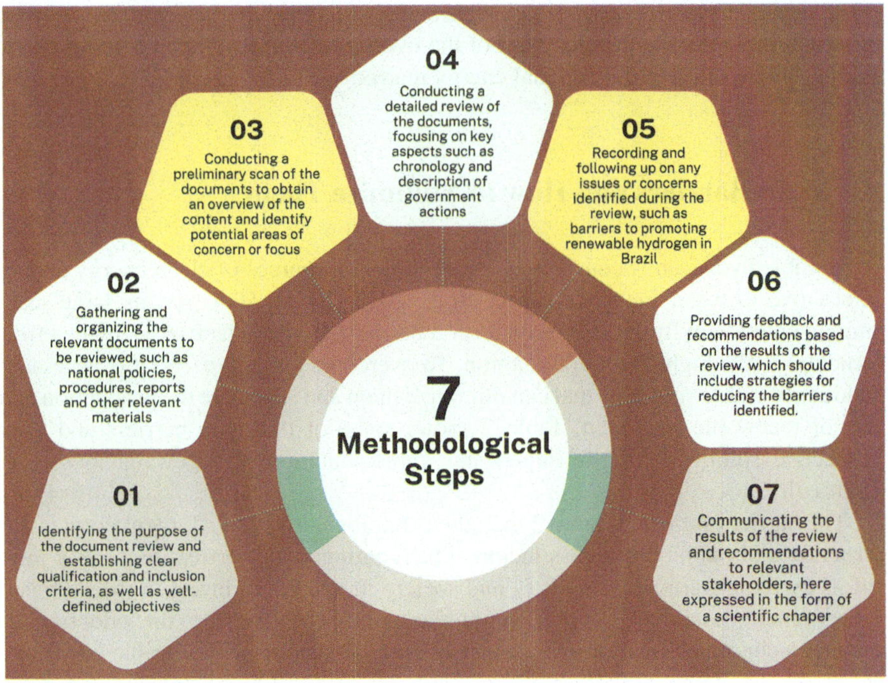

Fig. 2 Research methodology

particularly considering that a substantial portion—approximately 70%—of hydrogen production costs are intricately tied to electricity expenses. This economic landscape sets the stage for Brazil to potentially become a formidable hydrogen producer on the global stage, a prospect laden with transformative implications.

Forecasts suggesting Brazil's ascension to a prominent position in the hydrogen market signal a paradigm shift with far-reaching ramifications. The envisioned scenario holds the promise of catalyzing job creation, fostering innovation hubs, and beckoning substantial investment inflows. Moreover, it portends to reshape Brazil's commercial strategies, thereby redefining its role within the global value chain. The prognostication of Brazil's emergence as a pivotal hydrogen producer carries profound implications for the nation's trade dynamics, potentially exerting a positive influence on its trade balance. This transformative trajectory not only augurs well for Brazil's economic landscape but also bears testament to its pivotal role in steering the global transition toward a sustainable energy future.

However, large-scale production of renewable hydrogen requires significant investments in technology and infrastructure, which makes it necessary to create public policies to encourage its production and use (da Silva César et al., 2019). As such, Table 1 presents a compilation of the main government actions (programs, partnerships, roadmaps, plans, resolutions and decrees) that have been implemented in Brazil to encourage the large-scale production of renewable hydrogen.

Table 1 shows that the Brazilian government has been increasingly implementing measures to boost the deployment of renewable hydrogen in the country. Various programs, plans, partnerships, roadmaps, and resolutions have been created with the aim of stimulating the development of the hydrogen production chain and increasing its participation in the national energy matrix.

4 Main Barriers and How to Minimize Them

Renewable hydrogen is considered one of the main sources of clean energy and an alternative to fossil fuels (de Abreu et al., 2023, 2024). However, its large-scale implementation in Brazil still faces several barriers, from the lack of adequate infrastructure to the high cost of production. To overcome these barriers, it is necessary to identify viable solutions that can minimize them and stimulate the production and use of renewable hydrogen. Table 2 shows some of the main barriers and their respective solutions for the large-scale implementation of renewable hydrogen in Brazil.

The barriers shown in Table 2 include technological, financial, regulatory, infrastructure and public awareness factors. Overcoming these barriers requires a joint effort by the government, industry and society in general to invest in research and development, establish appropriate public policies, encourage the adoption of cleaner technologies and promote awareness of the benefits of renewable hydrogen. It should also be noted that all of this can be maximized by complying with the 5 Axes and their respective Guidelines of the PNH2 (MME, 2021).

Table 1 Policies in favor of renewable hydrogen

Initiative	Start period	Description	Reference
Brazilian Hydrogen and Fuel Cell Systems Program (Programa Brasileiro de Hidrogênio e Sistemas Células a Combustível—Procac, in Portuguese)	2002	This program aims to establish a national plan for the development of fuel cells, including hydrogen, in order to coordinate and conduct scientific and technological research, applied research and strategic research activities related to this topic	Chum (2002)
International Partnership for Hydrogen and Fuel Cells in the Economy (Parceria Internacional para Hidrogênio e Células Combustíveis na Economia—IPHE, in Portuguese)	2003	IPHE consists of international cooperation between the United States of America (USA) and several other countries, including Brazil, on hydrogen and fuel cell R&D, common codes and standards, and information sharing on infrastructure development	Portal Hidrogênio Verde (n.d.)
Science, Technology and Innovation Program for the Hydrogen Economy (Programa de Ciência, Tecnologia e Inovação para a Economia do Hidrogênio—ProH2, in Portuguese)	2005	Since 2005, Procac has been renamed ProH2, with the main aim of promoting initiatives that boost the advancement of hydrogen technology and fuel cell systems at a national level	MME (2021)
Roadmap for Structuring the Hydrogen Economy in Brazil	2005	The aim of this roadmap is to highlight the importance of various technological routes that offer competitive advantages for the country, such as water electrolysis, ethanol and other biomass. It also recognizes the role of natural gas in this transition and proposes a rationale for spreading the hydrogen market through distributed energy generation	Oliveira (2022)
RenovaBio Program	2016	The program aims to expand the production of biofuels (including biohydrogen), based on predictability and environmental, economic and social sustainability	EPE (n.d.)
Science, Technology and Innovation Plan for Renewable Energies and Biofuels 2018–2022	2018	The plan seeks to foster research, technological development and innovation in the energy production chains, including renewable hydrogen, to strengthen competitiveness and increase diversification of the energy matrix, ensuring energy security and efficiency	MCTIC (2018)

(continued)

Table 1 (continued)

Initiative	Start period	Description	Reference
National Energy Plan 2050 (Plano Nacional de Energia 2050—PNE, in Portuguese)	2020	The PNE, which is an instrument to support the design of the planner's long-term strategy for the expansion of the energy sector, provides for an increase in the share of renewable sources in the Brazilian energy matrix, including renewable hydrogen.	EPE (2020)
Resolution No. 2, of February 10, 2021, of the National Energy Policy Council—CNPE	11/02/2021	The resolution directs the National Electric Energy Agency and the National Petroleum, Natural Gas and Biofuels Agency to prioritize the allocation of research and development and innovation resources to the following topics: (i) hydrogen; (ii) nuclear energy; (iii) biofuels; (iv) energy storage; (v) technologies for sustainable thermoelectric generation; (vi) digital transformation; and (vii) strategic minerals for the energy sector	BRASIL (2021a)
Resolution No. 6, of April 20, 2021, of the National Energy Policy Council—CNPE	20/04/2021	This set out the guidelines for the National Hydrogen Program (PNH2) with a view to developing and consolidating the hydrogen market in Brazil and the country's international insertion on an economically competitive basis	BRASIL (2021b)
MME's National Hydrogen Program (Programa Nacional do Hidrogênio—PNH2, in Portuguese) do MME	2021	PNH2 sets out to define a series of actions that facilitate the joint development of three fundamental pillars for the successful progress of a hydrogen economy: public policies, technology and the market. The program's guidelines are structured around six axes, which include strengthening the scientific and technological bases, training human resources, energy planning, the legal and regulatory framework, opening up and growing the market, as well as competitiveness and international cooperation. The PNH2 guidelines indicate the paths that the national policy should follow, and the next stage will be the launch of the National Hydrogen Strategy, with the development of the guidelines.	BRASIL (2023) and Oliveira (2022)

(continued)

Table 1 (continued)

Initiative	Start period	Description	Reference
Energy Pact for Hydrogen	24/06/2021	The Brazilian government has launched the Energy Compact for Hydrogen as a voluntary commitment to promote the development of this market as a way of accelerating the reduction of GHG emissions. It was drawn up by a task force of the Ministry of Mines and Energy, the Ministry of Foreign Affairs and the Energy Research Company, with support from the Ministry of Science, Technology and Innovation, the Ministry of Education and other important partners. In the Pact presented, the Brazilian government committed itself to creating a knowledge base on H2 in Brazil based on three pillars, namely (i) research and development (R&D), (ii) capacity building and (iii) the creation of a collaborative platform called H2Brasil	EPE (2021)
COEMA Resolution No. 3 of 02/10/2022—1st Environmental Licensing Resolution for Green Hydrogen in the Country—State of Ceará	10/02/2022	The first environmental licensing resolution for renewable hydrogen in Brazil has been approved by the Ceará State Council (COEMA). Resolution No. 03/2022 sets out the procedures, criteria and parameters applicable to Environmental Licensing and Authorization within the scope of the State Environment Superintendence (SEMACE) for renewable hydrogen production projects for energy generation purposes in the state.	BRASIL (2022a)
Decree No. 11.075, of May 19, 2022	19/05/2022	The purpose of this decree is to establish the necessary procedures for drawing up Sectoral Climate Change Mitigation Plans, as well as set up the National System for Reducing Greenhouse Gas Emissions, which also involves expanding the use of renewable hydrogen. In addition, this decree also brings modifications to Decree No. 11.003, dated March 21, 2022	BRASIL (2022c)

(continued)

Table 1 (continued)

Initiative	Start period	Description	Reference
Resolution No. 6, of June 23, 2022, of the National Energy Policy Council—CNPE	23/06/2022	Establishes the National Hydrogen Program (PNH2) and establishes the program's governance structure. Before CNPE Resolution No. 6 of 2022, other decisions by the collegiate of ministers had already identified the strategic role that hydrogen can play in a future of decarbonized economies, such as CNPE Resolution No. 2 of 2021, which defined hydrogen as one of the priority themes for investments in research and development, and CNPE Resolution No. 6 of 2021, which guided the definition of guidelines for the program	BRASIL (2022b)
Bill of Law (PL) No. 725/2022	28/03/2022	The bill in question aims to include renewable hydrogen as an important energy source in Brazil, with a view to promoting a more sustainable energy transition. The proposal establishes parameters to encourage the use of sustainable hydrogen, seeking to create a favorable environment for the development of this industry in the country	BRASIL (2022d)

In addition to the solutions presented, it is important to highlight the need for a project monitoring system to maximize efforts to promote renewable hydrogen in Brazil. This system should monitor the progress of renewable hydrogen production, storage and distribution projects, as well as identifying possible bottlenecks and problems that need to be solved for these ventures to succeed. With proper monitoring, it is possible to improve resource management and reduce costs, as well as guaranteeing the safe and efficient use of renewable hydrogen. In this way, Brazil can consolidate its position as a leading player in the energy transition toward clean and sustainable sources.

5 Final Considerations

The energy transition in Brazil has become a vital issue for the country's sustainable development and requires special attention from companies, the government and society, among other stakeholders. In this context, renewable hydrogen presents itself as a promising opportunity for the national economy and can be used as an energy vector. However, in order to make this transition viable, it is necessary to

Table 2 Main barriers to large-scale implementation of renewable hydrogen and how to minimize them

Barrier	Description of barrier	How to minimize them?
Cost of production	Hydrogen production from renewable sources is still more expensive than production from fossil fuels, making renewable hydrogen less competitive on the market, which can discourage investment in large-scale production	To minimize this barrier, it is important to invest in research and development of technologies for renewable hydrogen production, with the aim of making it more efficient and less costly. In addition, the government can offer tax incentives and subsidies to companies that invest in clean technologies and reduce their GHG emissions
Infrastructure	The infrastructure needed for the production, storage and distribution of hydrogen is still in its infancy in Brazil, as well as in other countries with limited resources. New technologies and distribution networks need to be invested in to make large-scale production and use of hydrogen viable	To minimize this barrier, it is essential to invest in infrastructure for the production, storage and distribution of renewable hydrogen. This can be done through investments in research and development of storage and transportation technologies, as well as the construction of distribution networks that connect production sites to consumption sites. Another option is to establish public-private partnerships to invest in infrastructure, sharing the costs and risks between the government and the private sector
Regulations	The lack of specific regulations for the production and use of hydrogen can create legal uncertainty and hinder the development of the sector. Clear rules and tax incentives need to be established to stimulate the production and use of renewable hydrogen	To minimize this barrier, it is necessary to establish clear and consistent regulations for the production and use of renewable hydrogen, with the aim of providing legal certainty and encouraging the development of the sector. This can be done by creating specific laws, technical standards and regulations for hydrogen, which establish quality, safety and sustainability standards for the production, transportation and use of the fuel
Technical training	Hydrogen production involves advanced technologies and requires qualified professionals to operate and maintain them. It is necessary to invest in technical training and research and development to improve the efficiency and reduce the costs of renewable hydrogen production	To minimize this barrier, it is important to invest in technical training and research and development to improve the efficiency and reduce the costs of renewable hydrogen production. The government can create technical training programs to train qualified professionals in the production, operation and maintenance of hydrogen equipment and facilities

(continued)

Table 2 (continued)

Barrier	Description of barrier	How to minimize them?
Scale of production	Large-scale production is essential to reduce costs and make renewable hydrogen competitive on the market. However, there is still no significant demand for renewable hydrogen in Brazil, which can limit production capacity and the profitability of investments	To minimize this barrier, it is necessary to stimulate demand for renewable hydrogen in Brazil by creating public policies and incentives for the use of this fuel in various sectors. The government can create incentive programs for companies that use renewable hydrogen in their operations, such as subsidies for the purchase of equipment or tax exemptions
Availability of raw materials	The production of hydrogen from renewable sources depends on the availability of raw materials, such as water and biomass. The scarcity of these resources in some regions can limit the production of renewable hydrogen	To minimize this barrier, it is necessary to look for alternatives for the production of renewable hydrogen that do not depend on scarce resources in certain regions. One option is to invest in hydrogen production technologies from sources that are more abundant in Brazil, such as solar and wind energy
Public awareness and acceptance	The adoption of renewable hydrogen as an alternative energy source also depends on the awareness and acceptance of the general public, including in relation to the safety of the technology. The population needs to understand the benefits of renewable hydrogen and be willing to invest in technologies that use it	To minimize this barrier, it is necessary to invest in awareness campaigns and public education about the benefits of renewable hydrogen, including its positive environmental and economic impacts. In addition, it is important to ensure the availability of clear and accessible information on hydrogen production and use technologies, in order to clarify doubts and promote the adoption of renewable hydrogen as a sustainable alternative
Competition from other renewable energy sources	Renewable hydrogen competes with other renewable energy sources, such as solar and wind, which already have a more developed infrastructure in Brazil. This can make it difficult for renewable hydrogen to enter the energy market	To minimize this barrier, it is necessary to promote complementarity between the different renewable energy sources, recognizing the advantages of renewable hydrogen in terms of energy storage and transport. In addition, it is important to invest in research and development to improve the efficiency of renewable hydrogen production and reduce its costs, making it more competitive in relation to other renewable energy sources
Logistics	The production of renewable hydrogen can be located far from energy consumption centers, which makes it necessary to invest in logistics to transport the hydrogen to end consumers	To minimize this barrier, it is necessary to invest in transport infrastructure for hydrogen, such as gas pipelines and tanker trucks, in order to enable the distribution of renewable hydrogen on a large scale. In addition, decentralized hydrogen production can be encouraged in regions close to consumption centers, which would reduce the need for transport over long distances

(continued)

Table 2 (continued)

Barrier	Description of barrier	How to minimize them?
Technologies still under development	Some of the technologies needed to produce renewable hydrogen, such as fuel cells, are still under development and may take some time to reach the technological maturity needed for large-scale production	To minimize this barrier, it is necessary to invest in research and development to improve technologies related to the production of renewable hydrogen. In addition, encouraging partnerships between companies and universities can speed up the process of innovation and dissemination of technologies. It is also important to establish tax incentives and financing policies for companies that invest in research and development of technologies related to renewable hydrogen
Availability of funding	Large-scale production of renewable hydrogen requires high investments in infrastructure and technology. The lack of adequate national and international funding can limit companies' and governments' ability to invest in hydrogen production	To minimize this barrier, it is necessary to look for suitable sources of funding for the large-scale production of renewable hydrogen. This can include seeking out private investors and public-private partnerships, as well as tax incentives and government funding programs specifically for renewable energy projects. In addition, it is important that feasibility and financial return studies are carried out to attract investors and ensure the sustainability of renewable hydrogen production projects
Dependence on technology imports	Brazil is still dependent on importing technology for the production of renewable hydrogen, slowing down the development of the sector and increasing production costs	To minimize this barrier, it is necessary to invest in local research and development, encouraging the creation of startups and national companies that can develop innovative technologies and solutions. It is also important to establish partnerships with international companies and institutions to transfer knowledge and technology

Source: CGEE (2010), Da Silva César et al. (2019), MME (2021), Oliveira (2022), CNI (2022), Proença et al. (2023) and de Abreu et al. (2023, 2024)

monitor government policies aimed at increasing the use of renewable hydrogen, as well as the barriers and solutions for large-scale implementation.

Brazil has been implementing various actions to promote the use of hydrogen as a renewable and sustainable energy source. The Brazilian Hydrogen and Fuel Cell Systems Program (Procac), the International Partnership for Hydrogen and Fuel Cells in the Economy (IPHE) and the Science, Technology and Innovation Program for the Hydrogen Economy (ProH2) are some examples of government initiatives aimed at promoting renewable hydrogen. In addition, Bill of Law (PL) No. 725/2022, currently before the Chamber of Deputies, provides for the creation of a national program to encourage the use of renewable hydrogen as a clean and sustainable energy source.

It was possible to identify several challenges related to technological, financial, regulatory, infrastructure and public awareness aspects. To overcome these obstacles, there needs to be a joint effort between the government, industry, society and other stakeholders with a focus on investment in research and development, the implementation of effective public policies, encouraging the adoption of more sustainable technologies and raising awareness of the benefits of using renewable hydrogen. By implementing these actions, it is hoped that Brazil will be able to consolidate its leading position in the production and export of renewable hydrogen, while at the same time contributing to the transition to a more sustainable, low-carbon economy.

Given the need for further studies on the subject, it is recommended that research be carried out that takes a more in-depth look at the technical, economic and environmental feasibility of using renewable hydrogen on a large scale in Brazil. These studies must consider, for example, the potential for renewable hydrogen production in different regions of the country, as well as the infrastructure needed for its production, storage and transportation. In addition, it is also essential to assess the socioeconomic and environmental impacts of implementing a new hydrogen production chain, especially in terms of job creation, reducing GHG emissions and energy security. In this way, further studies could provide important input for improving public policies and advancing the use of renewable hydrogen in Brazil.

Acknowledgments This publication was made possible due to UFRJ's partnership with the H2Brasil project. The H2Brasil project is part of the German-Brazilian Cooperation for Sustainable Development and is implemented by the Deutsche Gesellschaft für Internationale Zusammenarbeit (GIZ) GmbH and the Ministry of Mines and Energy (MME) and funded by the German Federal Ministry for Economic Cooperation and Development (BMZ).

References

Bhandari, R., Trudewind, C. A., & Zapp, P. (2014). Life cycle assessment of hydrogen production via electrolysis–a review. *Journal of Cleaner Production, 85*, 151–163.

BRASIL. (2021a). *N° 5, de 11 de fevereiro de 2021. Resolução n° 2, de 10 de fevereiro de 2021, do Conselho Nacional de Política Energética – CNPE. Em 8 de março de 2021*. Disponível em: https://www.in.gov.br/en/web/dou/-/despacho-do-presidente-da-republica-307393461

BRASIL. (2021b). *N° 15, de 22 de abril de 2021. Resolução n° 6, de 20 de abril de 2021, do Conselho Nacional de Política Energética – CNPE. Em 14 de maio de 2021*. Disponível em:https://www.in.gov.br/en/web/dou/-/despacho-do-presidente-da-republica-320051164

BRASIL. (2022a). *Resolução COEMA N° 3 de 10/02/2022*. Disponível em: https://www.legisweb.com.br/legislacao/?id=427495

BRASIL. (2022b). *N° 63, de 1° de julho de 2022. Resolução n° 6, de 23 de junho de 2022, do Conselho Nacional de Política Energética – CNPE. Em 3 de agosto de 2022*. Disponível em: https://www.in.gov.br/en/web/dou/-/despacho-do-presidente-da-republica-419972141

BRASIL. (2022c). *Decreto N° 11.075, de 19 de Maio de 2022*. Disponível em: https://in.gov.br/web/dou/-/decreto-n-11.075-de-19-de-maio-de-2022-401425370

BRASIL. (2022d). *Projeto de Lei n° 725 de 2022*. Disponível em: https://www25.senado.leg.br/web/atividade/materias/-/materia/152413

BRASIL. (2023). *Programa Nacional de Hidrogênio – PNH2*. Disponível em: https://www.gov.br/mme/pt-br/programa-nacional-do-hidrogenio-1

Capurso, T., Stefanizzi, M., Torresi, M., & Camporeale, S. M. (2022). Perspective of the role of hydrogen in the 21st century energy transition. *Energy Conversion and Management, 251*, 114898.

Centro De Gestão E Estudos Estratégicos – CGEE. (2010). *Hidrogênio energético no Brasil. Subsídios para políticas de competitividade, 2010–2025, tecnologias críticas e sensíveis em setores prioritários*. Centro de Gestão e Estudos Estratégicos. Disponível em: https://www.cgee.org.br/documents/10195/734063/Hidrogenio_energetico_completo_22102010_9561.pdf/367532ec-43ca-4b4f-8162-acf8e5ad25dc?version=1.5

Chum, H. (2002). *Programa Brasileiro de Células a Combustível Proposta*. Disponível em: www.finep.gov.br/images/a-finep/fontes-de-orcamento/fundos-setoriais/ct-energ/programa-brasileiro-de-celulas-a-combustivel.pdf

Confederação Nacional Da Indústria – CNI. (2022). *Hidrogênio sustentável : perspectivas e potencial para a indústria brasileira*. Confederação Nacional da Indústria – CNI. 137 p.: il. ISBN 978-65-86075-53-3.

da Silva César, A., da Silva Veras, T., Mozer, T. S., & Conejero, M. A. (2019). Hydrogen productive chain in Brazil: An analysis of the competitiveness' drivers. *Journal of Cleaner Production, 207*, 751–763. https://doi.org/10.1016/j.jclepro.2018.09.157

de Abreu, V. H. S., Pereira, V. G. F., Proença, L. F. C., Toniolo, F. S., & Santos, A. S. (2023). A systematic study on techno-economic evaluation of hydrogen production. *Energies, 16*(18), 6542. https://doi.org/10.3390/en16186542

de Abreu, V. H. S., Da Ros Hollanda, D. L., Proença, L. F. C., Bahiense, L., & Santos, A. S. (2024). A systematic review on renewable hydrogen application in the land transportation sector. In *Intelligent transportation system and advanced technology* (pp. 9–31). Springer. https://doi.org/10.1007/978-981-97-0515-3_2

Empresa De Pesquisa Energética – EPE. (2020). *Plano Nacional de Energia – 2050*. Disponível em: https://www.epe.gov.br/pt/publicacoes-dados-abertos/publicacoes/Plano-Nacional-de-Energia-2050

Empresa De Pesquisa Energética – EPE. (2021). *MME apresenta ao CNPE proposta de diretrizes para o Programa Nacional do Hidrogênio (PNH2)*. Disponível em: https://www.epe.gov.br/pt/imprensa/noticias/mme-apresenta-ao-cnpe-proposta-de-diretrizes-para-o-programa-nacional-do-hidrogenio-pnh2-#:~:text=Em%20junho%2C%20o%20governo%20brasileiro,de%20gases%20de%20efeito%20estufa

Empresa De Pesquisa Energética – EPE. (n.d.). RenovaBio. Disponível em: https://www.epe.gov.br/pt/publicacoes-dados-abertos/publicacoes/renovabio#:~:text=O%20RenovaBio%20é%20uma%20iniciativa,sustentabilidade%20ambiental%2C%20econômica%20e%20social

Gulotta, T. M., Salomone, R., Lanuzza, F., Saija, G., Mondello, G., & Ioppolo, G. (2022). Life cycle assessment and life cycle costing of unitized regenerative fuel cell: A systematic review. *Environmental Impact Assessment Review, 92*, 106698.

Liu, W., Wan, Y., Xiong, Y., & Gao, P. (2022). Green hydrogen standard in China: Standard and evaluation of low-carbon hydrogen, clean hydrogen, and renewable hydrogen. *International Journal of Hydrogen Energy, 47*(58), 24584–24591.

Ministério Da Ciência, Tecnologia, Inovações E Comunicações – MCTIC. (2018). *Plano de Ciência, Tecnologia e Inovação para Energias Renováveis e Biocombustíveis*. Disponível em: https://antigo.mctic.gov.br/mctic/export/sites/institucional/tecnologia/tecnologiasSetoriais/Plano-de-Ciencia-Tecnologia-e-Inovacao-Para-Energias-Renovaveis-e-Biocombustiveis.pdf

Ministério De Minas E Energia – MME. (2021). *Programa Nacional do Hidrogênio. Posposta de diretrizes*. Disponível em: https://www.gov.br/mme/pt-br/assuntos/noticias/mme-apresenta-ao-cnpe-proposta-de-diretrizes-para-o-programa-nacional-do-hidrogenio-pnh2/HidrognioRelatriodiretrizes.pdf

Oliveira, R. C. (2022). *Panorama do hidrogênio no Brasil*. Disponível em: https://repositorio.ipea.gov.br/bitstream/11058/11291/1/td_2787_web.pdf

Portal Hidrogênio Verde. (n.d.). No Brasil. Políticas públicas. Disponível em: https://www.h2verdebrasil.com.br/no-brasil/

Proença, L. F. C., Coralli, A., Souza Toniolo, F., Darze, G. M., dos Santos Ribeiro Heluey, G., Martins, G. F., et al. (2023). Opportunities and challenges for the new hydrogen economy: Advances in renewable hydrogen. In *Transportation systems technology and integrated management* (pp. 121–140). Springer Nature.

Santos, A. S., de Abreu, V. H. S., de Assis, T. F., Ribeiro, S. K., & Ribeiro, G. M. (2021). An overview on costs of shifting to sustainable road transport: A challenge for cities worldwide. In *Carbon footprint case studies: Municipal solid waste management, sustainable road transport and carbon sequestration* (pp. 93–121). Springer.

Weidner, T., Tulus, V., & Guillén-Gosálbez, G. (2023). Environmental sustainability assessment of large-scale hydrogen production using prospective life cycle analysis. *International Journal of Hydrogen Energy, 48*(22), 8310–8327.

Zhang, J., Ling, B., He, Y., Zhu, Y., & Wang, Z. (2022). Life cycle assessment of three types of hydrogen production methods using solar energy. *International Journal of Hydrogen Energy, 47*(30), 14158–14168.

The Urban Energy Footprint Model

Anthony M. Yezer

1 Introduction

Household energy consumption in commuting and home use, heating and cooling, are substantial and vary across cities. Some causes of these differences, such as climate, are easily understood. Energy use for heating falls substantially with median January temperature and energy for cooling rises with July median temperature. Glaeser and Kahn (2010) find empirically that many city characteristics matter significantly. Gasoline consumption in commuting rises with the log of distance to the city center and falls with the log of population density of the residential area holding other factors such as income and household size constant. Their estimate suggests that average gasoline consumption per household falls by 20% when city size increases by 190%, or an elasticity of –0.1. The effect of city size on heating energy consumption is generally not substantial but location within the city has a significant effect. Suburban households consume substantially more heating and electrical energy than central city households presumably as a function of the lower density and larger size of housing units in the suburbs. Recently, Blaudin de Thé et al. (2021) have estimated gasoline consumption in French cities and found an inverse U-shaped relation in which consumption is low in both small towns and large cities but is the highest in moderate-sized cities. This is, in part a function of the widespread use of transit in large cities and the design of small towns. Thus, the empirical evidence indicates that a substantial amount of carbon emissions are based on commuting and residential real estate uses in cities and that the magnitude

A. M. Yezer (✉)
George Washington University, Washington, DC, USA
e-mail: yezer@gwu.edu

S. S. Muthu (ed.), *Energy Footprint and Sustainability*, Environmental Footprints and Eco-design of Products and Processes, https://doi.org/10.1007/978-3-031-63057-6_4

of these emissions varies substantially across cities even when accounting for differences in population. Informally, "urban sprawl" and congestion are clearly important.

While the empirical correlations in the literature relating energy use and carbon footprint to city characteristics are valuable, some theory is necessary to understand the causes of these differences and to understand the relation between changes in technology, behavior, or policy will can influence emissions in any given city. There was an existing model of urban land and housing markets, called the "standard urban model" (SUM), that was developed to understand the causes of sprawl and congestion in cities. The purpose of the SUM was to relate economic, technological, and policy-related differences to changes in urban form. Subsequently, Larson et al. (2012) adapted the SUM to generate the consequences of these differences in urban form for energy consumption in commuting and residences. This gave rise to the "urban energy footprint model" or the "UEF model."

Both the SUM and UEF models are capable of generating characteristics of the residential real estate in cities that should reflect the long-run equilibrium of cities. The models can be used to perform counterfactual exercises in positive economics. In such exercises, the change in a parameter reflecting technology, preferences, or policy for the configuration and emissions of the city is generated to reveal counterfactual alternatives. The SUM and UEF models can also operate normatively because the utility of city residents and rents of landowners are generated as model outputs. Therefore, the consequences of changes for overall social welfare can also be observed. A social planner could choose among these alternative outcomes the one which maximizes welfare.

The roadmap of this chapter begins with an introduction of the SUM, along with some interesting applications and classic results of the model. Readers may find some of these results quite counterintuitive. In the next section, the elements needed to convert the SUM into a UEF model, i.e., the addition of determinants of energy consumption, are discussed. One of these elements is the introduction of endogenous commuting congestion which both influences the form of the city and the level of energy used in commuting. In the third section, a series of specific applications of UEF models to help understand the implications of changes, both technological and policy related, for the energy footprint of cities are considered. Overall, the findings presented in the third section reveal two general points that complicate the analysis of the consequences of actions for the form and energy footprint of cities. First, the initial effects of a change are often offset by responses of urban systems that limit or even reverse the initial effects. Second, the long-run responses of city systems are very different than the short run. These two points are common in models of economic systems, and hence, their manifestation in urban models should not be surprising. The net result is that many of the results presented in the third section will appear counterintuitive.

2 Basics of Urban Economic Models

The urban energy footprint (UEF) model is based on the standard urban model (SUM) that originated with Alonso (1964) and was substantially extended by Muth (1985 and 1975) and Mills (1967). The model shows how an urban land market, operating without regulation, allocates urban land to housing in response to market forces and the urban transportation network, which is usually assumed to be exogenous. The distribution of employment in the urban area is usually given but what makes the model urban is that there is a concentration of employment in the central area that makes it necessary to import workers, i.e., commuting to the central business district (CBD) is necessary because of the employment concentration there. It is not necessary that all employment be located near the city center, only that the ratio of employment to resident population be substantially above unity near the center so that there is a flow of workers toward the center. Similarly, employment is generally more concentrated near the city center than population. The result of this difference in concentration is that there is net commuting toward the CBD.

In implementing the SUM to match a particular urban area, it is common to calibrate the model based on the actual spatial distribution of employment as a function of distance from the city center. This is easily accomplished by statistical estimation of an employment distribution function using approaches developed by McMillen (2004). Wheaton (2004) used the SUM to demonstrate that the spatial distribution of employment between the CBD and the rest of the city depends on a tradeoff between agglomeration economies that favor the CBD and congestion that raises the cost of CBD employment.

In addition to the assumption of exogenous employment distribution, the SUM assumes that workers are paid wages that are exogenous. Of course, in order to motivate commuters to travel to the CBD, wages vary inversely with distance from the city center. There are two versions of the SUM. In a "closed city" model, the population of the city is held constant. In order to maintain city population at a given level, either wages or worker utility in the city must vary. Alternatively, in "open city" models, workers must be indifferent between living anywhere in the city and residing in an alternative location where wages and cost of living are predetermined. In the open city model, city population is endogenous and utility is based on the exogenous level of utility available elsewhere as well as the exogenous wage in the city.

Another major feature of the SUM is that the construction of housing is based on formal models of a housing production function that uses land and structure inputs. Housing developers produce structures that provide housing services in a market that is generally regarded as competitive, i.e., there is an easy entry into the production of housing. However, housing production can be regulated as it is done by planning authorities in most cities. Households consume these housing services and the density of population is simply the quotient of housing space supplied in a given area and the quantity of space consumed by households living in that area. In essence, the characteristics of the residential real estate industry, including both supply and demand, are modeled and are fully endogenous to the model.

Given these assumptions about the concentration of employment and the need to commute, the original function of the SUM was to generate the long-run implications of changes in wages, transportation facilities, taxes, planning regulations, topography, etc., for the spatial distribution of housing and population. This means that the densities of residential structures and population in the city were generated by the model and then the long-run effects of changes in the circumstances of the city, including technology, and development policies, could be evaluated as counterfactuals. It is important to recognize that the model generates long-run counterfactuals, which in the case of cities could be 40 or more years because real estate investment depreciates slowly and the form of the city adjusts slowly to a new long-run equilibrium. Indeed, because of the long adjustment period and the speed of changes in building and transportation technology, consumer preferences, and public policy, the city is never observed in a long-run equilibrium. Precisely because the long-run effects of changes in conditions in a city are not apparent in the first few years after the change is implemented, the results of counterfactual analysis using the SUM are often quite surprising and counterintuitive. The contrast between short- and long-run effects of a change is often termed a "rebound" effect, and that term will be invoked frequently in this discussion of results obtained from using the SUM.

Perhaps the most well-known illustration of the difference between short- and long-run adjustments in cities is the effect of widening a highway on traffic congestion. In the short run, travel speed rises after the road is enhanced. In the long run, the rebound effect of automobile users may produce congestion that is as bad or worse than its initial level due to the response of housing and population to the transportation investment. The purpose of the SUM is to generate long-run counterfactual responses of the city to alternative changes rather than short-run effects. For this reason, results of experiments with the SUM are often counterintuitive. Short- and long-run changes in an urban system with significant time to build lags are often not only different in magnitude but even differ in sign.

In addition to providing counterfactual outcomes for cities undergoing different types of changes, the SUM is also able to produce welfare implications of these alternative outcomes. Wheaton (2004) demonstrated that the aggregate value of land is maximized in a welfare-maximizing city. Because the SUM has, among outcomes that it generates, aggregate land values, it is able to provide the results needed to rank counterfactual outcomes in terms of welfare. In a closed model, the change in wages needed to maintain worker utility and population can be used to compute changes in aggregate welfare. Finally, Bertaud and Brueckner (2005) have shown that welfare implications for a closed city with fixed population can be generated by observing the change in city limits, which reflects the additional commuting costs being imposed on workers. Thus, the SUM can not only create positive results, in which implications of changes in the city are generated, but it can also rank these outcomes in terms of implications for social welfare. In the jargon of economics, the SUM can be solved for outcomes that would be generated if a social planner directed the production of transportation facilities and housing to maximize welfare. Then those outcomes can be compared to counterfactual outcomes under alternative sets of policies or conditions.

A final question is how the assumptions made under the SUM compare to common notions of cities that are found outside economics. The answer is that there is remarkably close agreement. In the United States, the official notion of a city, a metropolitan statistical area (MSA), is based on a large densely inhabited core surrounded by areas of lower density where a significant fraction of workers commute to the core area. Specifically, MSAs require an urban area of 50,000 residents that is built up by appending census blocks which must have a density of at least 425 households per square mile. The urban area is built by appending census blocks that are sufficiently dense, specifically 200 households per square mile. The method allows for the inclusion of some blocks that have low housing density, called hops. In addition, the method allows the inclusion of land that has low density because of other nonhousing land uses, including industrial or commercial activity, i.e., airports. The urban area boundary can also "jump" over these nonresidential land uses and also over topographic features, particularly bodies of water, provided that densities on both sides of the impediment are sufficiently high.

Once an urban area reaches a total population of 50,000, it becomes a metropolitan statistical area (MSA). Then, counties are added to the MSA based on the work locations of individuals living in the subject county. If the fraction of workers in a county who are employed outside the county in the MSA is sufficiently large, that county is added to the MSA and the process of considering further counties is repeated. Clearly this emphasis on commuting to a central high-density core is consistent with the definition of a city in the SUM. Accordingly, the SUM is routinely applied empirically to MSAs. One caution is that some of the largest MSAs include secondary urban areas that can also be viewed as attractive for commuters. In this case, the alternative urban area is termed a primary metropolitan statistical area (PMSA). The combination of one or more PMSAs in a larger MSA forms a consolidated metropolitan statistical area (CMSA). This means some caution must be exercised in using the SUM to model an entire CMSA because it can have multiple dominant employment centers. However, there is a simple statistical test to determine whether the SUM can be applied to a CMSA with multiple large PMSAs. If a CMSA has a dominant central city, analysis of the spatial pattern of housing price or population density will show a single peak at the center. It is not necessary that prices decline at the same rate in all directions but the relation between commuting distance and population density must be monotonic and decreasing.

This discussion implies that the SUM may be used to model individual MSAs and draw conclusions about future effects on those areas of changes that are projected or proposed. This is certainly one use of the SUM. However, it is important to recall that the SUM gives long-run changes which may not be observed for 50 years because of the durability of existing real estate and the long time to rebuild at higher or lower density. An important use of the SUM is to answer generic questions about how a hypothetical city would respond to various changes, i.e., to generate general counterfactuals expected in a city of a given size.

3 Predictions of and Evidence for the Standard Urban Model

The foundation of the SUM is that, in the long run, households in a city will move whenever and wherever changes in location can raise their welfare. This is called an iso-utility condition and should not only hold within cities in a closed SUM but among cities in an open SUM. Of course, such arbitrage by moving from areas where welfare is low to high takes time and that is one reason, along with the substantial time to build real estate, why the SUM is a long-run equilibrium model. It is also important to recognize that, because responses of the city to changes are not observed quickly, the guidance of theory is necessary to distinguish short- from long-run effects.

Households are assumed to get utility from consuming housing, *h*, and other goods referred to as a composite commodity, *x*, whose price does not vary with location. Distance from the city center is measured by *k*. Households employed at the city center earn *y* and experience commuting costs of *tk* per time period. As noted above and tested empirically first by Madden (2009), some households are employed well outside the city center, i.e., at $k >> 0$ but earnings paid in this local employment, *y(k),* are lower than those at the city center because employers compete for workers with the city center and can pay a lower wage because their employees avoid commuting cost. Households maximize utility by choosing a composite nonhousing good whose price is normalized to unity everywhere to save notation and housing space per household, *h*, subject to a budget constraint

$$\begin{aligned} &\max_{c,h} U(c,h) \\ &s.t.\ \ y = c + tk + P(k)h \end{aligned} \tag{1}$$

where, as noted above, *t* is a uniform commuting cost per unit distance, *y* is exogenous income earned at the city center, and *P(k)* is the rental price of housing at distance *k* from the city center. Specifically, a necessary condition for households to maximize utility at the same level throughout the city is that

$$\partial P(k) / \partial k = -t / h \tag{2}$$

This is called Muth's equation, and it is a very important determinant of the structure of cities because it implies that, regardless of the conditions on the supply side of the housing market, the structure of housing prices over space is determined by the iso-utility condition of households and manifest by the structure of transportation costs, assumed to be a constant function of distance in the basic version of the SUM being discussed here.

Muth's equation, along with a few standard assumptions about the nature of the production and cost functions determining housing supply, generates a number of nontrivial testable hypotheses about the spatial structure of cities that are expected to hold provided that the urban land market is allowed to operate freely. One group of predictions concerns the general form of critical functions that determine the form of the city. The first of these functions is termed the housing rent function, $P(k)$, which is the relation between the rent per unit of housing space and distance. This is not the total rental cost of housing because that is the product of price per unit at k and the quantity consumed or $P(k)h(k)$. In the simplest terms, $P(k)$ is the rental price per square foot of housing adjusted for the quality of that space.

The second function is structure density, $S(k)$, which is the ratio of housing space to land and can be thought of as building density or height per unit land. Third is population density, $D(k)$, which is simply households per unit land. Fourth is the land rent function, $R(k)$, the rental price per unit land. Note that the housing price and land value functions are stated in terms of rental prices rather than asset prices because asset prices depend on interest rates, taxes, and other factors that can vary independently of rent over time and across cities. Some care must be taken not to confuse the sales prices of housing with the current rental price of that housing, which is less readily available or observed.

Structure density is based on a residential real estate production function taken from the literature. Builders combine structure inputs, whose price does not vary with location (or varies in a way that is determined exogenously by building regulations) with land. The general form of the housing production function is $h(k) = F(s(k), l)$, where h is housing services (usually measured by interior square feet of space), s is structure inputs, and l is buildable land used for the housing production at k. Given the rental price of housing, and the assumption that building is a perfectly competitive industry with zero economic profit, the builders determine an optimal structure density at each location, indicated by $S(k) = s/l$ ratio. The share of housing rents not required for structure inputs becomes the basis for the land rent, $R(k)$ function. Formally $R(k) = P(k)h(k) - p_s S(k)$. The iso-utility condition of households and iso-profit (at zero) condition of builders are crucial elements of the SUM.

The SUM predicts that these four functions, $P(k)$, $S(k)$, $D(k)$, and $R(k)$, are all nonlinear functions of distance and that, depending on the income elasticity of demand for housing, they can be approximated by negative exponentials as shown in Fig. 1. Note that the slope of the house price function is not steep and that the land rent function implied by that price function is very steep. All these functions have been subject to empirical testing in many academic papers covering different time periods and countries. The general result of this testing is that, provided that the land market is allowed to function and adjusting for limits on density due to city planning restrictions, the negative exponential functional forms implied by Fig. 1 fit cities remarkably well. The negative exponential house price function is even observed in cities where land use is heavily regulated. The most recent example of testing, performed on cities from around the world, can be found in Liotta et al. (2022). The results of this testing, in spite of the extreme difficulties of getting data on housing density and rents in much of the world, are remarkably consistent with expectations based on the SUM.

In addition to testing for the functional form and relative slope of the functions, the literature confirms a number of other predictions of the SUM regarding the form

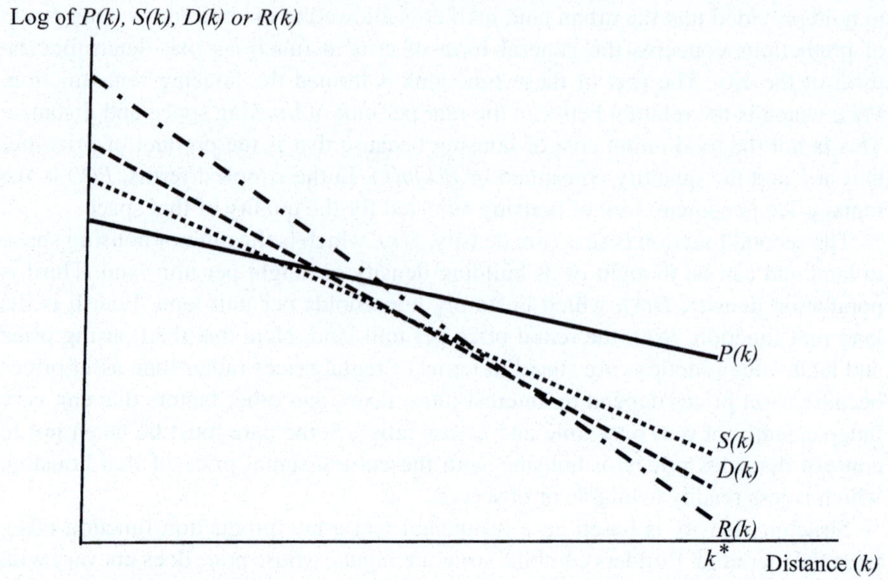

Fig. 1 Relations among housing rent, structure density, population density, and land rent gradients predicted by the SUM

and determinants of these functions. First note that Muth's equation, (2) above, predicts that the housing price gradient, the slope of $dP(k)/dk$, becomes flatter as income rises due to housing consumption increasing with income or as transportation costs fall. This result is confirmed in cross-sectional studies of cities, particularly comparisons across countries where differences in real income and transportation technology are pronounced. In addition, for countries experiencing substantial growth in real income per capita and falling transportation costs, like the United States, the slopes of all the gradients have fallen over time.

Another prediction of the SUM is that the slopes of the density functions do not vary with city size. This appears very counterintuitive because larger cities have higher prices, structure and population densities, and land rents at their centers. This increase in the intercept of the functions as distance approaches zero is consistent with the SUM. It is fairly easy to observe because city size and center city house prices, densities, and land rents clearly vary directly. Of course, this is trivially predicted by the SUM. However, the counterintuitive prediction of the SUM is about the slopes of the functions. Mills and Tan (1980) was the first of a number of empirical papers that tested the hypothesis that the slopes of one or more of these gradients are not a function of city size. This testing is generally performed using data for cross-sections of cities in various countries to determine whether the estimate of the gradient, usually a population density gradient, varies with city population. In all cases, the lack of a relation between city size and the slope of the gradient tested has been found to hold. A further test involves the relation between changes in the slopes of density gradients of cities over time and changes in income. The hypothesis is that the slope of the density gradient should fall with rising income, and this has also been confirmed.

This discussion leads to an important link between the SUM and the common notion of "urban sprawl," which is defined in many different ways outside economics. In terms of the SUM, sprawl is defined as a flattening of the slope of the population density gradient. For a city with a given topography, this means that the city boundary will increase for a given population. Topography is important because the normal pattern of city growth can be interrupted by topographic features. Approximately 40% of Chicago is interrupted by Lake Michigan. Granted that a small portion of the lake, north of Lake Street, was filled in as the city developed. However, the radius of the city would be far smaller for a given population if Lake Michigan did not exist. Measurement of sprawl in Chicago following the SUM is based on the falling population density gradient that has been driven over the last century by rising real income and falling transportation cost due to improvements in technology. These processes are important to understand because, as will be evident later in this chapter, falling density gradients or rising sprawl have significant effects on the energy footprint of a city but lowering real incomes or raising commuting costs are not popular as remedies for sprawl.

If a city is characterized by a constant population density gradient, then the growth rate in city population is mathematically identical to the growth rate of central density. This is apparent from the density formula. However, another implication of the SUM derived and tested by Zhao (2017) is the unitary elasticity property of cities, which states that even if the population density gradient is changing, as city population increases the sum of the elasticity of change in central density and the elasticity of land area with respect to population change is equal to unity. This property has been tested and found to hold for US cities.

An obvious prediction of the SUM based on Muth's equation is that rising transportation cost makes all of the four gradients in Fig. 1 steeper. In an early classic paper, Coulson and Engle (1987) find that differences in transportation costs steepen the housing price gradient and gasoline costs even appear to be overcapitalized in house values. This has been observed using a natural experiment in which spikes in gasoline prices in the 1970s and 1980s raised the cost of commuting by automobile substantially. Molloy and Shan (2013) estimated the effect of gasoline price shocks on the rate of urban housing construction as a function of commuting distance. They found that, for a large panel of cities, positive gasoline price shocks depressed the ratio of construction to housing stock when the average commuting time in the area was in the 24–28-minute time range. More recently, Larson and Zhao (2020) added another prediction of the SUM. They found that, in most cities, oil price increases raised the slope of population density gradients. However, this effect was reversed in the few cities that have petroleum as a major industry. In these cities, higher oil prices raise income and housing consumption and that more than counters the effect of gasoline price on commuting cost in Muth's equation.

Another very counterintuitive result of the SUM concerns what is termed the "urban wage premium." This is the elasticity of urban wages with respect to population, i.e., the percentage change in wages divided by the percentage change in city population. In computing percentage differences in wages by city, empirically great care must be taken to eliminate the effects of differences in education and experience associated with city size. The SUM not only predicts that the urban wage premium is not a function of city size but Jedwab et al. (2021) prove that strong versions of the SUM imply that, if the population density gradient is a negative exponential, that the urban wage premium is given by

$$d\log y \,/\, d\log N = \beta\eta \tag{3}$$

where y is earnings per worker adjusted for differences in age, education, etc., N is total city population, β is the fraction of household income spent on rent generally calibrated to equal 0.3, and η is the share of land in the cost of residential housing (calibrated to equal 0.2). Based on this calibration, the urban wage premium in the United States is $\beta\eta = 0.06$. This coincides almost exactly with empirical estimates of the urban wage premium, carefully adjusted for differences in worker skill, reported separately in Chauvin et al. (2017) and Liu (2017). In sum, both a strong version of the SUM and empirical evidence suggest that a 20% increase in total city population is associated with a 1.2% rise in wages paid to workers holding their education, experience, and demographic characteristics constant. Testing has also confirmed this relation for other countries. It provides a strong empirical confirmation of the predictions of the SUM.

A final counterintuitive implication of the SUM is the proposition that the population and employment density functions move independently. Put another way, sprawl of population does not cause sprawl of jobs and sprawl of jobs does not cause sprawl of population. Of course, both jobs and population tend to sprawl as transportation cost falls but these moves are independent of one another. Thurston and Yezer (1994) tested this hypothesis of the SUM for US cities and found it to hold. Furthermore, they tested the effects of center city crime and racial composition on changes in the slope of the density gradient over time and found that these factors were not significant in promoting sprawl. Note that the SUM requires that testing for sprawl or decentralization be based on changes over time in the slopes of the employment and population density gradients using a Granger causality test. Other testing that is not based on the SUM or any formal model may yield different results.

This review of implications and tests of the SUM, when the theory is applied to actual cities, has been designed both to establish the scientific merit of the model and illustrate the counterintuitive nature of many of its predictions. The SUM also draws attention to a different set of measures of the city, specifically to various density and price gradients in Fig. 1. Adding energy use in housing and transportation, as discussed in the next sections, to form the energy footprint of the city model increases its complexity and accordingly raises the prevalence of counterintuitive counterfactual results.

4 Adding Energy Use in Commuting to the Standard Urban Model

Most commuting in cities is by automobile, and the common approach in urban energy footprint (UEF) models has been to assume that commuting is done by private automobiles powered by internal combustion engines traveling on roads that have no tolls or fees for use. More recently, these assumptions have been relaxed substantially but the initial UEFs were based on automobile commuting. Adding energy use by automobile to the SUM involves two interactive steps, one straightforward and the other complex. First is the determination of the relation between vehicle operating conditions and energy use. Second is the estimation of the operating conditions given the tendency for roads to become congested and vehicle velocity to slow substantially. The results of the model are easily modified to deal with changes in vehicle fuel efficiency, highway tolls, high-occupancy requirements, and electric vehicles. Indeed, examples of such modifications to the baseline automobile commuting case will be discussed in one of the sections of this chapter.

Because fuel consumption is based on commuting velocity, it is necessary for the energy footprint model to consider the effects of congestion on vehicle velocity and fuel efficiency. West et al. (1999) conducted an automobile fuel efficiency study at the Fuels, Engines, and Emissions Research Center at the Oak Ridge National Laboratory. The study was based on observations of the performance of a weighted average of the gasoline-powered vehicles tested. The resulting data are used to parameterize the relationship between velocity and gasoline consumption for a representative vehicle in a simulated city. A good example is the miles per gallon equation estimated by Larson et al. (2012), which is a fourth-degree polynomial in velocity:

$$M_i = \alpha + \sum_{j=1}^{4} \beta_j V_{ij} + \epsilon_i \tag{4}$$

where M is miles per gallon, V_{ij} is vehicle velocity raised to the power j, and the βs are parameters to be estimated. The parameter estimates of Eq. (4) have been used in the urban energy footprint (UEF) model to estimate gasoline consumption as a function of vehicle velocity. The fuel efficiency function is based on gasoline with 100% petroleum content. Peak fuel efficiency in miles per gallon is reached at about 40 miles per hour. This is below the vehicle velocity in the outer suburbs but far greater than the speeds in inner-city congestion. An alternative to basing fuel consumption on average velocity would be to include the possibility that traffic congestion involves stopping and starting. This approach would be possible, but it has not been implemented in the literature.

The second challenge is to determine the velocity of commuting vehicles as a function of distance, k, from the city center. Velocity depends on congestion, which is a function of vehicle traffic per unit land used for roadways. Commuters using the roadway at a given distance, k', are determined by the difference between cumulative employment from $k = 0$ to k' and cumulative households living at $k \leq k'$. If the model assumes all employment is in the city center, then this computation depends only on the population density function. McMillen (2004) provides a good example of how to estimate and understand urban density functions. Formally employment located inside distance k' is given by the integral of the employment density function from 0 to k'.

The population commuting on highways at k' is the difference between the integrals of the employment density and population density functions. The population density function in the SUM is determined endogenously in the model as discussed below.

Finally, the relation between commuting volume at k', highway capacity at k', and vehicle velocity must be determined. This relation, which has been developed by traffic engineers, is taken as given by the SUM and was developed by Muth (1975) and O'Sullivan (1985). Commuting velocity at a given location is inversely related to traffic volume according to

$$v(k) = \frac{1}{a + bM(k)^c} \tag{5}$$

where $v(k)$ is the commuting speed at distance k from the CBD, $M(k)$ is the ratio of traffic volume to road service through location, k and a, b, c are parameters that are calibrated to the specific city being modeled to reflect the severity of the traffic congestion function. Calibration of the a, b, and c parameters is accomplished by observing velocity of travel at different points in the city. Using this information along with values of $M(k)$ at those points, the parameters of this nonlinear relation can be calibrated. This particular formulation and calibration approach was first used by Muth (1975) and subsequently used in many efforts to understand the spatial form of the city because congestion is an important determinant of vehicle velocity.

Velocity is not only important to the SUM because of its effects on fuel consumption, which is part of the cost of commuting, but it is most important because vehicle velocity determines the time cost of commuting. Time and operating costs must be considered separately because fuel efficiency and gasoline tax policies can change one without changing the other. Obviously, time cost is related to wages, which are a fundamental factor in determining urban form and city size. Of course, vehicle velocity at a given distance is a deeply endogenous variable in the SUM because it is determined by the spatial pattern of endogenous residential location. Clearly, the interactions among these variables are complex and can easily lead to counterintuitive counterfactual simulation results.

The land used for streets is generally assumed to be a constant fraction of available land, usually set at 25%. This fraction is based on Bartholomew (1955) and Muth (1975). However, the model can be solved for situations in which the fraction of land for streets is a function of k. Clearly, this can be a policy variable. The SUM can also be solved with tolls on the roads that effectively "price" the marginal congestion cost experienced by commuters. Clearly transportation is crucial for understanding the city because, if transportation costs were zero, there would be no city because, as Muth's equation shows, the housing price gradient would be flat.

Given the explanation of vehicle velocity discussed above, the total commuting cost from distance *k'* to the CBD is given by

$$T(k) = (\theta w)\int_{k=0}^{k'}\left(1/v(k)dk\right) + \tau\int_{k=0}^{k'}\left(1/M\left(v(k)\right)dk\right)\tau k' \tag{6}$$

where θw is the value of time spent commuting expressed as a fraction of the wage, w, and τ is the cost of fuel and operation of the vehicle which allows the inclusion of gasoline or carbon taxes. The marginal cost of commuting used in Muth's equation is $dT(k)/dk$ in a UEF model rather than the constant t found in ordinary versions of the SUM. Clearly in an energy footprint model, this derivative is not constant due to the effects of congestion on velocity of travel. This substantially complicates solution of the model and requires numerical techniques rather than closed-form analytical solutions.

Finally, the energy and carbon use implications of commuting follow in a fairly straightforward fashion. The total fuel consumed by commuters from *k'* in Eq. (6) is given by $\tau\int_{k=0}^{k'}(1/M(v(k))\,dk)$, and the energy content and carbon emissions are a function of the nature of the fuel being consumed. In the case of hybrid or fully electric vehicles, this requires assumptions about the nature of the power source producing the electricity. These issues will be covered in the discussion of specific applications of the UEF model below.

5 Adding Energy Use by Households to the Standard Urban Model

Energy use by households in the SUM is complicated because of the many different choices of housing types in the city and the direct and indirect implications of housing types for energy consumption. Most energy use is based on climate control, i.e., heating and cooling the structure. This depends on climate, and some assumption about temperatures in the city must be made at the start. The next important determinant of climate control cost is the number of exterior walls in the structure. The SUM has nothing to say directly about this issue. But it does predict the density of housing structure, i.e., the structure/land ratio. The common practice since Larson et al. (2012) has been to relate the density generated by the SUM to structure types, multifamily, single family attached, and single family detached. The structure/land ratios from the SUM are matched to actual ratios of these different housing types from the city to generate a baseline distribution of structure types that matches the observed city.

The indirect implication of housing choice for energy use is based on appliance choice by the household. There is a well-developed empirical literature on the choice of appliances by households assuming that housing choice, including both the type of unit and its location, is predetermined. Empirical analysis of the simultaneous choices of housing type and location and appliances has proved elusive in part because of the additional complication that energy costs of operating appliances vary with location. However, a primary justification for the approach commonly used in UEF models is provided by the Housing Assistance Supply Experiment (HASE), which was conducted to determine what factors influence household energy consumption. Neels (1982) analyzed the experimental data and concluded that changing the structure type from single family detached to multifamily lowers energy use by 20–50% compared to 5–10% reductions from alterations due to energy efficiency upgrades of a given structure type and 2–5% from changes in occupant behavior due to variations in energy price structures. In sum, the effects of structure type dominate the determinants of differential energy consumption in the HASE data, which have the advantage of being generated by a randomized control experiment.

As a result of these complications, household energy use in the UEF model assumes that housing unit size is an exogenous determinant of energy use by households. The estimation of this energy demand equation follows logically from the work by Ewing and Fang (2008) and Glaeser and Kahn (2010). Using data from the Residential Energy Consumption Survey for various years, it is possible to estimate household energy use equations of the form

$$\ln E_i = \alpha + \beta \ln A_i + \lambda \ln I_i + \pi \ln p_i + \Sigma_j \gamma_j S_{ij} + \omega X_i + \varepsilon_i \tag{7}$$

where E is the energy consumption of household i, A is the area of interior space in the dwelling, I is the income, p is the average price paid per thousand BTUs, the S_{ij} are dummy variables describing the structure type, and X is a vector of other controls such as heating and cooling degree days that are commonly used in the literature. Estimates of (7) using ordinary least squares are performed under the assumption that the energy use of individual households does not influence the characteristics of the housing unit. Inclusion of structure characteristics and climate variables allows the estimates of the equation to be adapted to the differential circumstances in various cities where influences of climate and building material on energy use are consequential.

The UEF model generates a structure density function, $S(k)$. This is mapped into a series of structure types, multifamily, single family attached, and single family detached, that correspond to the S_j categories in the Residential Energy Consumption Survey data. Thus, the continuous $S(k)$ function produces a series of structure types that are a stepwise linear function of k. As a result, the household energy use simulations from the model are piecewise linear. This is a consequence of the assumption that, at any given distance, all of the housing has the same structure type.

As was the case in adding commuting energy use to the SUM to form the UEF model, household energy use may be elaborated for various purposes. For example, in order to evaluate policies that change energy prices, it is necessary to make energy cost a separate component of the rental price of housing. As a result, raising energy cost provides an incentive to lower the energy consumption per square foot of housing as well as to consume smaller housing units. These complications will be covered in relation to specific applications of the model considered below.

6 Solving the UEF Model Version of the SUM

For purposes of this chapter, the UEF model is discussed as a special version of the SUM because it includes three features. First, energy use in commuting. Second, household energy use. Third, congestible commuting. This third feature is crucial because congestion has a substantial effect on energy consumption but mainly because endogenous highway congestion complicates solution of the model and eliminates the possibility of analytical solutions.

One method of model solution is discussed here. It follows the approach adopted in Larson et al. (2012), which was the first explicit version of the UEF model. The model solution is achieved by computing the solution to a two-equation system of differential equations. Given the assumptions regarding the demand for housing services and the technology of housing production as well as fixed household income in the CBD, the SUM allows solution for the population density function, *D(T(k))*, in terms of commuting cost. The total population housed within distance *k* of the CBD is the integral of the population density function, $N(k) = 2\pi\mu\int_{k=0}^{k} k\,D(T(k))dk$, where μ is the fraction of land used for housing and *D(T(k))* is the population density function. The number of households housed at distance *k* is given by

$$\frac{dN(k)}{dk} = 2\pi kD(T(k)) \tag{8}$$

Recognizing from the discussion above that the *v(k)* function depends on *N(k)*, Eqs. (6) and (8) imply a simultaneous relation between commuting costs and population distribution that must be solved as a system of first-order differential equations.

Fortunately, the initial values for population and transportation costs at the edge of the CBD are known. Because no population is housed inside the CBD and employment inside the CBD is a known constant, the number of commuters to the CBD at its edge is known to be E_{CBD}. Highway capacity at the edge of the CBD is also known and equal to *H(0)*. Velocity and travel time can be computed as *v(0)* and *t(0)* based on *H(0)* and E_{CBD}. The system of differential equations with initial values to be solved can be written as

$$\frac{dT(k)}{dk} = m_1 + \frac{p_G}{G(v(V(E_L(k), N(k), H(k))))} + \frac{w}{2v(V(E_L(k), N(k), H(k)))} \tag{9}$$

$$\frac{dN(k)}{dk} = 2\pi\theta\, kD\big(T(k)\big) \tag{10}$$

with initial values

$$T(0) = m_0 + k_{CBD}\{m_1 + \frac{p_G}{G(v\big(V\big(E_L^T,0,H(0)\big)\big)} + \frac{w_i}{2\,v(V\big(E_L^T,0,H(0)\big)} \tag{11}$$

$$N(0) = 0 \tag{12}$$

where E_L^T is the total local employment in the city and *N(0) = 0* because no households live within the CBD. The earlier discussion established that the distribution of local employment is predetermined along with the spatial distribution of highway capacity. Thus, there are two covariates, *N(k)* and *T(k)*, in the system of ordinary differential Eqs. (9) and (10) with known initial values at the edge of the CBD.

There is no explicit solution to this system of differential equations due to the complexity of the functional forms. Instead, the system is solved numerically. Once the commuting cost and number of households living in each annulus of the city are computed, it is possible to solve for all other variables that characterize that annulus of the city because, as is generally the case with the SUM, the housing price, housing consumption per household, and housing production at each distance are determined by commuting cost from that location due to the iso-utility and iso-profit conditions discussed above.

More recent versions of the UEF model discussed in subsequent sections of this chapter have added complications including alternatives to automobile travel. These require households at each *k* to choose a mode, and this modal choice alters the highway congestion function which intern alters *T(k)*. However, the basic form of the solution algorithm remains the same and the reader can consult these individual papers, many of which are reviewed below, regarding the nuances of the alternative solution techniques.

7 Examples of and Results from the UEF Models

This section presents the examples chosen to illustrate the diversity and flexibility with which the SUM and UEF models can be applied to issues related to energy use in cities. Also, some important insights regarding extensions of the approach that can be gained from the current literature are identified. It is important to remember that formal UEF models have only been studied for just over a decade and the potential insights that the UEF approach could offer are largely untapped. This section is divided into subtopics that deal with the different questions that have been addressed in the UEF model literature.

7.1 *The SUM and UEF Models and Optimal Pricing of Automobile Commuting*

Even before the advent of the SUM, the problems arising from the failure to price highways were well known in economics. The analysis of the problem, called a "market failure," is straightforward. Commuters decide to use highways based on the private cost to them, which appears as the *T(k)* function in the discussion above. This is the marginal private cost of the trip. But the total social cost of the trip is the sum of private cost and the additional cost imposed on others. The nature of this external additional cost, called a congestion externality, can be seen from Eq. (5) by recognizing that the time needed to travel one unit distance is *(1/v),* and if *M(k)* is written simply as the ratio of commuters, N, to roadway space, R, at *k,* then the effect of an additional commuter on travel time per unit distance is

$$d\left(1/v\right)/dN = bc\left(N^{c-1}\right)/R^{c} > 0 \qquad (13)$$

However, this increased travel time is imposed on all *N* commuters so that the total travel time delay or externality due to one additional commuter at *k* is $N[d(1/v)/dN] = bc(N/R)^c > 0$, which is an increasing function of *N.* Consequently, the congestion externality associated with the marginal commuter in cities can be substantial particularly near the city center where the number of commuters is very large. For a detailed discussion of the issue of congestion pricing and the related problem of optimal investment that preceded the study of the issue in an urban spatial context, see Keeler and Small (1977). The application of these principles to the UEF model is straightforward.

Given that household location decisions are based on private commuting cost which excludes the extra component of social cost in (13), the actual spatial pattern of households in cities is unlikely to be optimal. This congestion externality problem has a substantial relation to the energy footprint of cities because it suggests a resource allocation problem that exists independent of the issue of energy use or carbon footprint.

Even before concerns with the UEF model, the implications of pricing automobile transportation in cities to reflect the external cost in (13) attracted substantial attention. Indeed, this problem inspired the Muth (1975) initial simulation model with congestion. Obviously raising transportation cost makes the slopes of all the gradients steeper and results in a city with higher central density and shorter city radius holding population constant. Since then, there have been several notable attempts to study the effects of pricing automobile commuting externalities.

The first detailed exploration of the effects of congestion pricing in an urban economic context was done by Wheaton (1977). Comparing a simulation of actual cities with a counterfactual where the congestion externality was priced, he found that central residential densities would be four times as high as the actual city. However, the model assumes that all employment is located in the CBD. Some care must be taken to set the optimal congestion toll. First, the toll is equal to the marginal congestion cost imposed on other commuters based on Eq. (13). However, the toll revenue must then be rebated, along with land rent generated by the housing sector to city residents; otherwise, welfare would fall due to the effects on resident income. Thus, the welfare-maximizing toll is determined in a model where toll revenue is returned equally to each city resident in a closed city and the welfare-maximizing toll minimizes the wages that must be paid to city workers in order for them to achieve the same level of utility in the city before and after the toll is imposed. The sum of the fall in aggregate wages plus any rise in land rent gives the rise in welfare associated with the toll. Proque et al. (2020) have applied the UEFM to analysis of the effects of urban transportation policy in Brazilian cities.

Perhaps the most precise simulation of the difference between an actual city and one with congestion pricing is McDonald (2009). The model was calibrated to Chicago in 1954 because that year saw the Chicago Area Transportation Survey, a massive study of commuting that divided the city into seven concentric rings. The survey revealed that about 30% of all commuting trips terminate in the CBD and that there was no commuting from the CBD to the outer ring. Indeed, the most common destination ring for commuters from the seventh ring of the city was to the CBD. Also, 74% of all commuting trips were taken during rush hour, generating substantial congestion. All this is consistent with assumptions of the SUM and explains why Chicago has been so heavily studied in the SUM and UEF modeling literature. Because the survey gives a detailed origination and destination information on trips and the distribution of streets is known, the degree of congestion throughout the city with no congestion charges is observed. The SUM allows the population density function to be simulated with and without congestion pricing. The function without congestion pricing is $\ln D(k) = 4.541–0.150\ k$ while with congestion pricing it is $\ln D(k) = 4.762–0.178\ k$. This means that the central density increases by *0.221* log points or an increase of *23.7%*, i.e., a rise from *93.8* to *117* households per acre. Given the higher central density, in order to hold the population constant, the slope of the density gradient increased and the city radius fell. Congestion pricing substantially lowered sprawl.

While these exercises were performed with the SUM and energy implications were not generated, it is obvious that commuting cost fell due to shorter commutes and residential energy consumption should also fall as structure density would have to rise to accommodate the higher central densities. What is most consequential about such results is that they show a policy failure, i.e., government failing to price traffic congestion, resulting in a substantial increase in energy use in cities. Of course, eliminating the market failure actually raises aggregate land rent and that rise in rent, with no falling wages, and holding population constant, implies that the policy actually raises welfare even before considering the effects on energy use or carbon emissions.

Paradoxically, these results, generated before the first formal UEF models were implemented, point toward what is likely the most efficient way to reduce the energy footprint of cities, i.e., price congestion in automobile commuting. Congestion pricing has the important advantage that it enhances general urban efficiency by allocating use of highways to users based on equating marginal social cost of the travel to marginal benefit to the user. The consequent fall in carbon footprint due to greater housing density and lower commuting distances is an extra benefit of this welfare increase. Put another way, having the public sector price highway access not only generates welfare gains, it also has the potential to significantly lower energy consumption in cities.

There is one limitation to this conclusion regarding the effects of congestion pricing of automobile travel in cities on energy consumption. The versions of the UEF model discussed thus far in this section assume that commuting is done by individual workers. Recently, Zhao (2020) formulated a version of the UEF model in which there is possible carpooling in response to the cost of operating a single vehicle. The main purpose of this model is to understand the energy consequences of high-occupancy vehicle (HOV) lanes, which will be discussed in more detail in a subsequent section. However, it has implications for the literature on congestion pricing discussed in this section and illustrates an important point about modeling the complex forces determining energy use in cities.

Zhao (2020) makes the number of commuters per automobile endogenous. Her approach is straightforward and illustrates the flexibility available under the current versions of UEF models. Specifically, individual commuters are assumed to commute the cost of commuting as a function of the number of commuters per vehicle. The assumption is that commuters will divide the operating costs of commuting equally. Obviously, the time cost is not divided and each commuter is assumed to need to pay the fixed costs for an automobile so that only the operating costs are divided. Alternatively, each commuter drives on a different day. The number of commuters per car is determined by the lowest cost alternative. Importantly, there is a time cost imposed to collect commuters so that adding commuters raises time cost and lowers operating cost. In the case of tolled roads considered here, the toll payments are another operating cost. Note that congestion tolls vary with distance based on road crowding and that is the case here. Clearly, the cost-effectiveness or advantage of carpooling rises with distance because operating cost plus tolls rise with distance. In addition, carpooling lowers the number of vehicles on the road and raises vehicle velocity, which lowers the operating cost. The result is that the number of commuters per vehicle is an increasing function of the length of the commute. Residents living close to the CBD still drive alone.

Allowing carpooling produces a rebound effect that offsets the tendency of congestion tolls to raise city density, lower city radius, and save energy used in both commuting and housing due to the higher housing price encouraging smaller units and more dense construction. Note that carpooling does not change the fact that congestion pricing raises aggregate welfare. Indeed, allowing carpooling enhances welfare. However, because it lowers both the operating cost per rider and raises vehicle velocity, allowing carpooling substantially flattens the housing price gradient. The result that the combination of congestion tolls and carpooling results in a city radius that is slightly larger than the city without either of these two innovations. Energy use in housing rises but carpooling lowers energy use in commuting substantially. Overall, the carbon footprint of the city still falls while the welfare gain from tolls is even larger when carpooling is allowed.

This example illustrates some very important points about the use and interpretation of the UEF models. There are many margins on which households can adjust to changes in incentives to live and work in the city. When one of these margins, carpooling incentives in this example, is ignored, the consequences for counterfactual estimates of the effects of changes on energy consumption can be substantial. In this case, the initial effects of congestion tolls, even in the long run, are all in the direction of lower carbon footprints. Commuting distance is shorter and energy use in commuting falls. Housing prices rise, which lowers household energy consumption both because consumers choose smaller units and because the density of housing unit construction rises. Introduction of carpooling allows a rebound effect in which the relation between distance and commuting cost is changed. Specifically, this relation is based on the lower bound of commuting cost with one, two, or three commuters per car, introducing substantial additional nonlinearity in the housing price function. As a result, the city radius can actually increase and housing density falls slightly as average housing consumption rises. This provides a very useful case example of the way in which adding complexity to the UEF model can modify or in this case even introduce an important rebound effect.

7.2 *The UEF Model and Taxation to Lower Emissions in Cities*

There are many proposals to use tax policy to lower greenhouse gas emissions, generally lumped under the heading of "carbon taxation." The implications of such taxation for society are difficult to model because of the substantial diversity of effects. For example, climate has a substantial effect, through heating and cooling, on the energy consumption of households. Glaeser and Kahn (2010) computed very substantial differences in the carbon tax implications for cities in temperate climates, San Diego, versus those in either cold, Boston, or hot, Miami, climates. Clearly, such substantial geographical differences in tax burdens would have major implications for the growth and decline of individual cities. Of course, these effects could be mitigated if the tax revenue were to be rebated to the residents of individual cities based on the average tax revenue collected.

The UEF model allows investigation of the effects of carbon taxes on individual cities. It also allows consideration of substitution of alternative taxes that could achieve similar outcomes for city carbon footprints. Constructing a version of the UEF model that can achieve these ends requires some additional modifications to the versions considered thus far.

Important efforts to analyze the effects of taxation on the form of and emissions by cities were provided by Bento et al. (2006) and determine the effects of development taxes in Bento et al. (2011). However, the most ambitious attempt to add tax policy to the UEF model is by Borck and Brueckner (2018). This work is discussed at some length here. As is often the case with models involving the imposition of taxes and subsidies, the goal of this model is to generate the welfare implications of different tax strategies and determine the welfare effects of alternatives that achieve the same changes in carbon emissions. The "social planner's problem" is to select tax strategies that reduce emissions to an aggregate level that is below that in an unregulated city and accomplish this reduction in a fashion that imposes the smallest cost on society. This was done by using a closed city version of the SUM so that city population is held constant.

The planner's problem is to minimize the cost in terms of wages that must be paid to residents to occupy the city subject to the constraints that city population is constant, the utility of residents is unchanged, and emissions are reduced to a predetermined level. Logically if the same number of residents are willing to live in the city and accept lower wages, then the collective welfare of society is higher because the city has been made more attractive. This approach assumes that all cities are identical and identical tax policies are implemented everywhere. If cities are not identical or taxation policies vary, then utility levels in other cities will change and population will move. The problem of modeling the effects of tax and other carbon policies on cities of different sizes has not been considered in the UEF model literature. However, it is clear that the policies being considered, particularly tax policies, are not spatially neutral in their effects.

Given that the imposition of various taxes on residents of a city reduces welfare, it is necessary that the closed UEF model include a rebate of tax revenue to residents. This is done in what is termed a "lump sum transfer" so that the incomes of all residents are raised regardless of their location within the city. Rebating tax revenue tends to attenuate effects on energy consumption because higher income raises housing consumption, but this is necessary in order to measure the welfare effects of alternative policies.

A number of additions to or modifications of the UEF model structure not discussed earlier are needed for the consideration of tax strategies. First, it is necessary to generate separately each variable that can be used as the base of a tax. If gasoline is to be taxed, then usage of gasoline must be modeled. This, of course, is a trivial modification of the standard UEF model which considers gasoline usage as a function of commuting distance.

Second, if energy use in housing is to be taxed separately, the relation between housing consumption and household energy must be established. One approach would be simply to make energy use in housing a function of housing space, h(k), and to add this cost to the rental price of housing space. However, this ignores the possibility that producers of housing can alter the energy efficiency of housing space through changes in technology or construction methods. Clearly, these opportunities are important as using alternative fuels, heat pumps, or better insulation are important adaptations of housing space in reaction to higher energy prices. The result is that energy use in housing is modeled in a far more elaborate manner that is unique to this chapter. Housing structures are squares occupying land. Continuing to follow Borck and Brueckner (2018), the area of the sides and roof of the structure is given by

$$4fh(s)(l)^{0.5} + l \tag{14}$$

where l is the land area occupied by the structure and $fh(s)$ is the height of one side with f transforming height based on s inputs into a physical height of the building side. Energy use per unit of building surface area is then given by $e\ \{4fh(s)(l)^{0.5} + l\}$, where e is a parameter reflecting the rate of energy loss through surfaces. Then this means that the energy use per unit land can be written as

$$e\left\{\left[4fh(s)/(l)^{0.5}\right]/l\right\} = Ah(s)e + e \tag{15}$$

where $A = 4f/l^{0.5}$. Equation (15) is very convenient because total energy use in housing is given by the integral of this expression over land used for housing between the CBD and city boundary.

Because developers have to consider the cost of structure inputs and energy, they consider the rental price of housing net of energy cost given by $P(k) - A\ p_e\ eh(s)$, where p_e is the price of energy that can include taxes. This netting out of energy cost from the rental price of housing means that a second-order effect of the home energy tax is to lower structure density which actually raises energy consumption in the city. But the higher cost of housing lowers housing unit sizes. Thus, the possibilities for complex rebound effects in this model abound. Land rent determination requires that the cost of both structure inputs and energy be deducted before land rent is determined.

$$R(k) = P(k) - Ap_e eh(s) - p_s S(k) - p_e e \tag{16}$$

The consequent reduction in land rent for housing tends to shift the $R(k)$ function down and result in a smaller city radius as housing is now less competitive with agricultural land at the edge of the city. The explicit modeling of household energy consumption adds a significant element of extra complexity to the UEF model.

Third, it is necessary to attribute some harm to total greenhouse gas emissions from the city. This could be done by simply adding a reduction to household income based on the aggregate harm attributed to the aggregate emissions in the city. In this case, the harm from emissions is included in the utility function so that the household's problem from Eq. (1) is now written as

$$\begin{aligned} &\max_{c,h} U(c,h,G) \\ &s.t.\ y = c + tk + P(k)h(s) + Ap_e eh(s) + p_e e \end{aligned} \tag{17}$$

where G is the total greenhouse gas emissions from the city, including those related to gasoline consumption in commuting and household emissions. In this case, the marginal utilities of the composite commodity, c, and housing, h, are positive and that of greenhouse gas emissions, G, is negative, i.e., $\partial U/\partial c > 0$, $\partial U/\partial h > 0$, and $\partial U/\partial G < 0$. In essence, greenhouse gasses are being treated as what is termed a local public bad, i.e., as if they were a general pollutant. The harm done by the emissions does not depend on the density of population in the area where they originate so that it is not necessary to model *G(k)*.

Given this model structure, it is possible to impose a variety of taxes on the city and determine the levels of these taxes that solve the social planner's problem to produce a welfare maximum. Borck and Brueckner (2018) consider explicitly: taxing floor space at rate τ_h to address energy use to due to heat transfer through the sides of the building, a land tax, τ_l, which is related to energy loss through the building roof, and a commuting tax per mile of τ_t that is justified based on the carbon footprint of automobile commuting. Note that this model does not have congestion, and, hence, computation of this tax is relatively straightforward as vehicle velocity is constant throughout the city. These are all specific or excise rather than ad valorem taxes. That is. the land tax is levied at a uniform amount per unit land throughout the city, i.e., it is relatively trivial near the CBD and quite consequential near the city edge. Also, the tax on building density is uniform regardless of the price of housing. The commuting tax is based on the quantity of gasoline used regardless of the net of tax gasoline price. One insight of the model is that the form of taxation needed to deal with greenhouse gas emissions efficiently is rather different than the taxes commonly levied, except for the gasoline tax.

The results of the model show that achieving the planner's goal of a limited amount of aggregate emissions, *G,* is possible using these three taxes. However, when only one of the three instruments is available, the rate on that tax must be set exceedingly high. When all three are employed, relatively modest rates are sufficient to achieve the welfare-maximizing reduction to achieve the aggregate emissions goal. For each of these counterfactual experiments, the model generates actual solutions for the changes in the form of the city that occur when the emissions goal is achieved. In all cases, the solutions involve both reduction in commuting and housing size as the city limit falls, population density rises, and housing consumption per household falls. House prices rise near the city center as do land rents.

This model is unique in that it shows that the UEF model can be used to analyze the effects of carbon tax policies on cities and demonstrate the levels of taxation needed to achieve emissions targets while maintaining household utility.

7.3 The UEF Model and High-Occupancy Vehicle (HOV) Lanes in Cities

High-occupancy vehicle (HOV) lanes are reserved for vehicles with multiple passengers. One motivation for reserving some lanes for HOV travel is the reduction of congestion during peak commuting periods. There is substantial literature about the circumstances in which the provision of HOV lanes is likely to lower congestion. The difficulty is that, in diverting some lanes from ordinary to HOV use, congestion is increased on the ordinary lanes. In addition, the fall in travel time on HOV lanes induces more travel. The net result can be slower average travel times overall. These issues are explored in some detail in Hughes and Kaffine (2019) as well as several other studies and the case for HOV lanes, based on partial equilibrium analysis of consequences for travel times, is complex.

The UEF model considers reserving some portion of highway capacity for autos with two or more occupants in a general equilibrium context in which the total number of potential commuters within the city is held constant but their residential locations are allowed to vary. Zhao (2020) recently developed a method to add HOV lanes to the UEF model that is the basis for the discussion provided here. This is a closed city model, but the utility of city residents is allowed to vary with earnings held constant. This form of the UEF model is used for positive analysis of the effects of policy changes rather than normative social planning applications.

The clever insight into the model is that the provision of HOV lanes provides an alternative to single driving that has differentiated cost functions over space. Specifically, adding a second rider to the automobile allows the commuters to split the variable costs of driving but not the time cost. There is an additional time cost associated with each additional rider added to a carpool. The net cost of commuting is then the lower bound of commuting cost assuming one, two, or three riders in the vehicle. As distance increases, the advantages of adding riders increase and there are switch points where carpooling begins and where additional riders are added to the carpool.

The relative cost of carpooling depends on the fraction of highway capacity allocated to HOV lanes. However, the general effect of HOV lanes as discussed earlier is that the number of vehicles falls and vehicle velocity generally increases throughout the city. Both of these effects lower energy consumption. However, the fall in commuting cost is most pronounced as distance from the center increases, which means that the housing rent gradient flattens, the city sprawls, and housing space consumption increases. The surprising result is that this rebound effect of increasing energy consumed in housing virtually completely offsets the fall in commuting energy. Overall, the energy footprint implications of converting some highway capacity to HOV use are negligible.

7.4 *The UEF Model and Implications of Work from Home*

Work from home (WFH), which may be "telework," has been increasing rapidly, partially in response to public health concerns. Descriptive empirical evidence shows that WFH increases as a function of commuting time to the employer. It is also true that WFH rises with the fraction of workers in management, business, scientific, and arts occupations.

Telework effects have been analyzed using the SUM. Safirova (2006) allows a fraction of WHF in the city under the assumption that these workers do not contribute to the agglomeration economies of their employers. The result is that there is an equilibrium in which the growth of the CBD is limited by WFH in the suburbs. The consequences for energy consumption and the carbon footprint of the city are not part of this model.

As part of the general theme with issues considered here, the partial equilibrium effects of work from home (WFH) appear to lower carbon emissions as the amount of gasoline burned in commuting falls for two reasons. First, there are fewer automobiles; and second, vehicle velocity rises as congestion falls. But there is a substantial rebound effect that is apparent when effects on spatial location of households in the city are considered. Clearly, as the fraction of days spent in WFH rises, commuting cost falls far more than the congestion reduction. Furthermore, the saving in commuting cost raises income available for consumption of the composite commodity and housing.

All of these effects have been incorporated into a UEF model by Larson and Zhao (2017). In this model, the authors consider carefully the carbon intensity of energy used in commuting, housing, and consumption of the composite goods. They find that as the fraction of work time involving WFH rises, the city sprawls and high-density housing declines by approximately 50%. Allowing one day per week of WFH with no reduction in pay raises household utility by 3.2%, increases land used per structure by 7.4%, and interior space of units by 4.0%. As a result, energy consumed in housing rises by 5.3%, but commuting energy consumption falls by 23%. The net effect is a 0.4% rise in energy consumption because the aggregate amount of household energy consumption is larger than the energy used in commuting. Finally, the implications for changes in greenhouse gas emissions are negligible because the carbon content of commuting energy is larger than household energy, which includes substantial electrical energy. Thus, the rebound effect of rising household emissions just balances the initial fall in commuting emissions. Of course, the ultimate effect on the form of the city, particularly as days per week in WFH rise, is very consequential. There is substantial sprawl and growth in housing consumption.

The model can also consider other possibilities. For example, what happens if the benefits of WFH are captured by employers or competed away so that wages fall enough to eliminate the rise in utility. This reduces the increase in housing consumption and results in a slight fall in the carbon footprint of the city. At this point, the effects of WFH on worker productivity and wages have not yet been determined but the UEF model can accommodate any possibility.

7.5 Adding Mass Transit to the UEF Model

Mass transit is an alternative to automobile commuting that has been neglected in the review of applications of UEF models thus far. Adding mass transit to the model is not difficult because it is entered as an alternative to the automobile with its own commuting cost function. There are two alternative ways in which transit options have been added to the UEF models. Larson and Yezer (2015) show that increasing city size has little effect on carbon footprints per capita as long as cities rely entirely on automobiles for commuting. But city size and density allow more mass transit alternatives.

Borck (2019) assumes mass transit is a general alternative available as a function of distance with its own travel time and cost based on system characteristics. Households at any distance are then confronted with two alternatives whose cost differs. The fraction of all households choosing mass transit is based on the relative cost of the two modes. As transit cost rises compared to the automobile, the fraction choosing the auto rises toward unity. Transit has a lower carbon footprint than the automobile. Hence, the direct effect of introducing transit is to lower carbon emissions. However, there is a rebound effect as providing transit, particularly if it is attractive, lowers commuting cost, which encourages sprawl. This raises emissions. Overall, the net result of introducing a transit alternative is a small fall in emissions. Lowering fares after the introduction of transit lowers emissions further. However, because this model lacks endogenous congestion of the roads, the potential for transit to lower automobile commuting cost is neglected.

Agarwal and Zhao (2023) model the introduction of both mass transit by fixed rail and ride-hailing in a city. This model is particularly advanced and computationally challenging because transit routes radiate out from the CBD, which matches reality. This means that selected corridors of the city have excellent transit service while others do not. In order to model this, the authors effectively slice the city into wedges with different levels of service and analyze the effects of commuting cost on housing separately in each wedge. Ride-hailing is added as a service that can either provide commuting to the center or short trips to mass transit. This research shows the potential of the UEF model to consider the effects of multimodal transit provision in a realistic setting. It is perhaps the most advanced and computationally challenging of all UEF models implemented thus far.

Overall, the results of these efforts indicate the effects of adding mass transit to a city on its carbon footprint are modest even when the substantial energy implications of installing mass transit systems are ignored.

8 Summary and Conclusions

The UEF model is slightly more than a decade old. The approach has been applied to a variety of changes ranging from HOV lanes to telework to determine effects on both urban form and carbon footprints. Similarly, alternative policies for transportation, planning regulation, and taxation have been considered. For example, Larson et al. (2022) find that zoning regulations have little effect on the energy footprints of cities. The examples presented here are a small fraction of the potential applications of the UEF model.

An advantage of UEF models is that they consider household utility effects. This means that they can be used in normative analysis in which the welfare consequences of alternative proposed changes are evaluated. Using the UEF model in the social planner mode is very promising.

One common theme of all the applications presented in this chapter is that responses of urban real estate markets to changes are complex and often counterintuitive. This makes checking for rebound effects and perverse policy implications with such models particularly important. The examples presented in this chapter are only representative of the future possibilities of analysis using UEF models.

One result of the UEF model does seem particularly strong and attractive. A policy that increases the welfare of city residents and lowers the carbon footprint of cities is to charge congestion tolls for the use of city roads provided that the toll revenue is rebated locally to residents. Unfortunately, this policy, which could be implemented electronically through fast-pass technology, has not even been considered in the menu of possible approaches to lowering carbon emissions because the connection between these tolls and urban housing markets is not intuitive. This example best illustrates the power of the UEF model to differentiate among alternative strategies for lowering energy consumption and greenhouse gas emissions in cities and identify first-best policies.

References

Agarwal, D. R., & Zhao, W. (2023). Taxing Uber. *Journal of Public Economics, 221*, 1–25.

Alonso, W. (1964). *Location and land use: Toward a general theory of land rent*. Harvard University Press.

Bartholomew, H. (1955). *Land use in American cities*. Harvard University Press.

Bento, A., Franco, S. F., & Kaffine, D. (2006). The efficiency and distributional impacts of different anti-sprawl policies. *Journal of Urban Economics, 59*(1), 121–141.

Bento, A., Franco, S. F., & Kaffine, D. (2011). Is there a double dividend from anti-sprawl policies. *Journal of Environmental Economics and Management, 61*(2), 135–152.

Bertaud, A., & Brueckner, J. K. (2005). Analyzing building-height restrictions: Predicted impacts and welfare costs. *Regional Science and Urban Economics, 35*(2), 109–125.

Blaudin de The, C., Carantino, B., & Laforcaude, M. (2021). The carbon "Carprint" of urbanization: Evidence from French cities. *Regional Science and Urban Economics, 89*, 1–28.

Borck, R. (2019). Public transit and urban pollution. *Regional Science and Urban Economics, 77*, 356–366.

Borck, R., & Brueckner, J. (2018). Optimal energy taxation in cities. *Journal of the Association of Environmental and Resource Economists, 5*(2), 481–516.

Chauvin, J. P., Glaeser, E., Ma, Y., & Tobio, K. (2017). What is different about urbanization in rich and poor countries? Cities in Brazil, China, India and the United States. *Journal of Urban Economics, 98*(1), 17–49.

Coulson, E. N., & Engle, R. F. (1987). Transportation costs and the rent gradient. *Journal of Urban Economics, 21*, 287–297.

Ewing, R., & Fang, R. (2008). The impact of urban form on U.S. Residential Energy Use. *Housing Policy Debate, 19*(1), 1–30.

Glaeser, E. L., & Kahn, M. E. (2010). The greenness of cities: Carbon dioxide emissions and urban development. *Journal of Urban Economics, 67*(3), 404–418.

Hughes, J. E., & Kaffine, D. (2019). When should drivers be encouraged to Carpool in HOV Lanes. *Economic Inquiry, 57*(1), 667–684.

Jedwab, R., Loungani, P., & Yezer, A. (2021). Comparing cities in developed and developing countries: Population, land area, building height, and crowding. *Regional Science and Urban Economics, 86*(1), 1–18.

Keeler, T., & Small, K. (1977). Optimal Peak load pricing and investment and service levels on urban expressways. *Journal of Political Economy, 85*(1), 1–15.

Larson, W., & Yezer, A. (2015). The energy implications of city size and density. *Journal of Urban Economics, 90*(1), 35–49.

Larson, W., & Zhao, W. (2017). Telework: Urban form, energy consumption, and greenhouse gas implications. *Economic Inquiry, 55*(2), 714–735.

Larson, W., & Zhao, W. (2020). Oil prices and housing demand. *Real Estate Economics, 48*(3), 8098–8849.

Larson, W., Liu, F., & Yezer, A. (2012). The energy footprint of the city: Effects of urban land use and transportation policies. *Journal of Urban Economics, 72*(2–3), 147–159.

Larson, W., Yezer, A., & Zhao, W. (2022). Urban planning policies and the cost of living in large cities. *Regional Science and Urban Economics, 96*, 1–18.

Liotta, C., Viguie, V., & Lepetit, Q. (2022). Testing the monocentric standard urban model in a global sample of cities. *Regional Science and Urban Economics, 97*(1), 1–19.

Liu, S. (2017). Agglomeration, urban wage premiums, and college majors. *Journal of Regional Science, 57*(4), 611–630.

Madden, J. F. (2009). Urban wage gradients: Empirical evidence. *Journal of Urban Economics, 18*, 291–301.

McDonald, J. F. (2009). Calibration of a monocentric urban model with mixed land use and congestion. *Regional Science and Urban Economics, 39*(1), 90–96.

McMillen, D. P. (2004). Employment densities, spatial autocorrelation, and subcenters in large metropolitan areas. *Journal of Regional Science, 44*(2), 225–244.

Mills, E. S. (1967). An aggregative model of resource allocation in a metropolitan area. *American Economic Review, 57*(2), 197–210.

Mills, E. S., & Tan, J. P. (1980). A comparison of urban density functions in developed and developing countries. *Urban Studies, 17*(3), 313–321.

Molloy, R., & Shan, H. (2013). The effect of gasoline prices on household location. *The Review of Economics and Statistics, 94*(4), 1212–1221.

Muth, R. F. (1975). *Cities and housing: The spatial pattern of urban residential land use*. University of Chicago Press.

Muth, R. F. (1985). Numerical solution of urban residential land use models. *Journal of Urban Economics, 2*(4), 307–332.

Neels, K. (1982). Reducing energy consumption in housing: An assessment of alternatives. *International Regional Science Review, 7*(1), 69–81.

Proque, A. L., dos Santos, G. F., Betarelli Jr, A. A., & Larson, W. D. (2020). Effects of land use and transportation policies on the spatial distribution of urban energy consumption in Brazil. *Energy Economics, 90*, 1–12.

Safirova, E. (2006). Telecommuting, transportation, and agglomeration: A general equilibrium model. *Journal of Urban Economics, 52*, 26–52.

Sullivan, A. (1985). The general equilibrium effects of the residential property tax: Exercise and excess burden. *Journal of Urban Economics, 18*(2), 235–250.

Thurston, L., & Yezer, A. M. (1994). Causality in the suburbanization of population and employment. *Journal of Urban Economics, 35*(1), 105–118.

West, B. H., McGill, R. N., & Sluder, S. (1999). *Development and verification of light-duty modal emissions and fuel consumption values for traffic models* (Report FHWA-RD-99-068). U.S. Department of Transportation, Federal Highway Administration.

Wheaton, W. C. (1977). Residential decentralization, land rents, and the benefits of urban transportation investment. *American Economic Review, 67*(2), 138–143.

Wheaton, W. C. (2004). Commuting, congestion, and employment dispersal in cities with mixed land use. *Journal of Urban Economics, 55*(3), 417–438.

Zhao, W. (2017). The unitary population density property in a monocentric city with a negative exponential population density gradient. *Regional Science and Urban Economics, 62*, 1–11.

Zhao, W. (2020). The general equilibrium effects of high-occupancy vehicle lanes on congestion, sprawl, energy use, and emissions. *Journal of Regional Science, 60*, 174–200.

Exploring the Intersection of Human Psychology, Business Analysis, and Project Management to Reduce Computer Equipment's Environmental Impact

Richard Govada Joshua

1 Introduction

Our digital footprint's environmental impact is becoming a more urgent problem at a time when technology is being relentlessly integrated into every aspect of our life. The complex dynamics at the intersection of human psychology, project management techniques, business analysis, and the energy footprint of computer equipment are explored in this chapter. By analyzing how individual and organizational decisions impact the environmental impact of information technology (IT), it aims to disentangle the intricate interactions between these factors and their combined influence on energy consumption within the IT business.

The decisions we make in navigating the digital realm, both personally and within our organizations, have a significant impact on how energy-efficient computer equipment is. Using behavioral psychology as a source of inspiration, this chapter aims to clarify the complex cognitive processes and decision-making variables that affect energy use in the IT industry (Lauesen, 2019). In the pursuit of environmentally friendly information technology practices, it delves beyond the traditional examination of hardware and software to investigate the human element, which is sometimes disregarded.

A key component of this investigation is the use of a solid business analytical framework and project management perspective. Through the application of business analysis techniques, this chapter takes care to identify and analyze the key factors that impact energy efficiency (Walker, 2015). This analytical tool not only offers a thorough grasp of the current situation, but it also lays the groundwork for

R. G. Joshua (✉)
Comprobase Inc., Sterling, VA, USA

New York Institute of Technology, New York, NY, USA
e-mail: joshuarichardgovada@ieee.org

S. S. Muthu (ed.), *Energy Footprint and Sustainability*, Environmental Footprints and Eco-design of Products and Processes, https://doi.org/10.1007/978-3-031-63057-6_5

the development of strategic methods to lessen the environmental impact of computer equipment.

Additionally, this chapter takes a project management perspective, acknowledging the difficulties involved in the shift to sustainable practices. By highlighting the value of collaboration and adaptability, it successfully negotiates the complexities of putting environmentally responsible initiatives into practice. The chapter tackles the real-world obstacles of implementing environmentally friendly IT practices by integrating project management concepts into sustainability projects.

Essentially, the perspective that this chapter promotes is holistic—that is, one that crosses conventional lines and unites the disparate disciplines of project management, business strategy, and human psychology. Through the adoption of this comprehensive viewpoint, it hopes to make a significant contribution to the current discussion on environmental sustainability in the IT sector. As a result, it not only emphasizes the importance of solving the ecological difficulties brought by our digital achievements, but it also provides realistic insights that might drive the sector toward a more sustainable and responsible future.

2 Contribution

1. Investigating and Understanding Energy Footprints in Computing.
   Approaching the complicated interaction of technology and human behavior from a multidisciplinary perspective, efforts are made to comprehend computing's energy footprint according to Fig. 1. Through an examination of the interactions between the energy footprint of computer equipment and project management approaches, business analysis, and human psychology, researchers can obtain a deeper understanding of the complex dynamics that shape patterns of energy usage (Lauesen, 2019). While a strong business analytic framework identifies the critical elements driving energy efficiency, behavioral psychology sheds light on how organizational and individual decision-making affects energy usage within the IT industry (Tversky & Kahneman, 1974). The foundation for well-informed plans to solve environmental concerns is laid by historical assessments, which offer evidence-based insights into the evolution of energy footprints in computers (Kahneman, 2011).

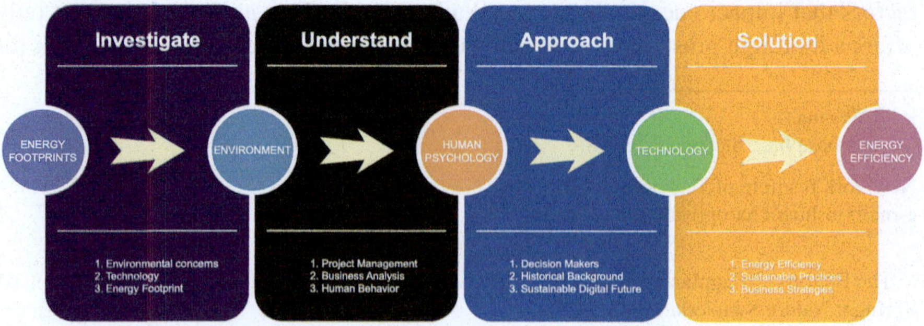

Fig. 1 Evolution of energy footprints in computing

2. Applying Solutions and Approaches to Address Environmental Challenges.
   Strategic actions informed by interdisciplinary perspectives are necessary to address environmental concerns in the IT sector. According to Fig. 1, using knowledge from project management, business strategies, and behavioral psychology, workable methods for reducing environmental impact and improving energy efficiency are revealed. The suggested methods emphasize collaboration and adaptability as essential elements and are intended to promote sustainable practices in the sector (Kahneman, 2011). Also, the identification and execution of concrete steps to lessen environmental effect are made possible by useful approaches based on the principles of business analysis (Lauesen, 2019). These initiatives support the adoption of more environmentally sustainable computing methods in the future by promoting holistic viewpoints and taking historical contexts into account.

3 Investigating and Understanding Energy Footprints in Computing

3.1 Investigating the Intricate Interplay Between Computer Equipment's Energy Footprint and Human Psychology, Business Analysis, and Project Management Approaches

In the fast-changing environment of information technology (IT), the energy footprint of computer equipment has arisen as a key problem, leading a look into its complicated interplay with human psychology, business analysis, and project management methods (Tversky & Kahneman, 1974). According to Fig. 1, using insights from behavioral psychology, business analytical frameworks, and project management methodologies, this investigation tries to uncover major determinants impacting energy efficiency and suggests tactical approaches to reduce environmental footprints (Thaler & Sunstein, 2008) (Fig. 2).

1. Human Psychology and Decision-Making.
   Understanding the relationship between human psychology and computer equipment's energy footprint is critical to developing successful energy-saving solutions (Lauesen, 2019). Behavioral psychology provides useful information about how organizational and individual decision-making processes influence energy consumption patterns. For example, the concept of cognitive biases influences energy-related decisions significantly (Tversky & Kahneman, 1974). The tendency to emphasize short-term benefits above long-term sustainability might result in inefficient energy use habits. Cognitive biases have a large impact on energy-related decision-making (Thaler & Sunstein, 2008). When information on energy use is presented, the anchoring effect—where people rely on initial information as a reference point—can influence decisions toward more sustain-

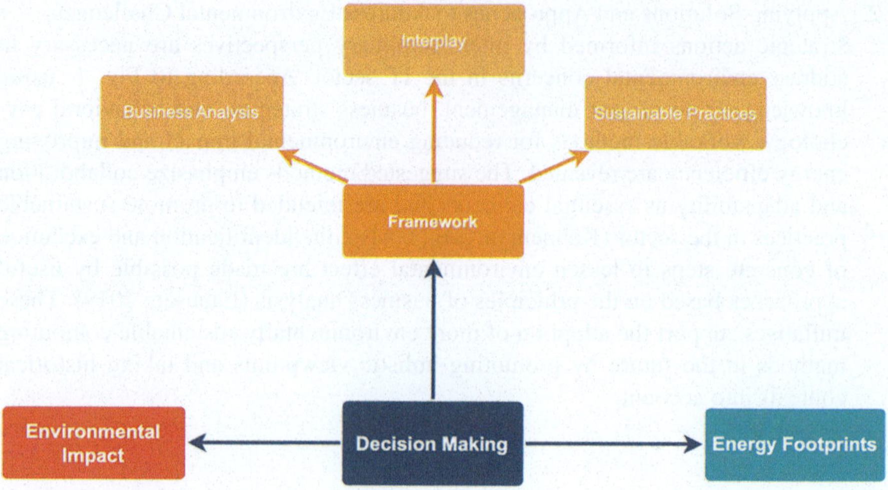

Fig. 2 Incorporating human psychology, business analysis, and project management

able habits. Furthermore, confirmation bias, which is defined as the tendency to seek information that confirms current opinions, can be strategically addressed by offering facts and data on the environmental impact of alternative energy options. This strategy acts as a mild nudge, directing decision-makers to environmentally beneficial solutions (Walker, 2015). Furthermore, the current bias, characterized by a preference for short-term advantages over long-term benefits, may lead to decisions that prioritize immediate convenience over energy-saving activities. According to psychological studies, incentives can have a major impact on behavior (Kolk & Pinkse, 2005). Financial incentives, recognition, and gamification can all be used to encourage people and businesses to adopt energy-efficient activities (Hawken, 2017). Positive reinforcement, such as recognizing and praising energy-saving efforts, can establish a positive feedback loop that supports long-term decision-making.

2. Business Analysis Frameworks.

   Using business analysis frameworks in the context of energy efficiency offers a methodical way to pinpoint and deal with important variables. The business analysis lens allows for a thorough study of processes, stakeholders, and key performance indicators (KPIs) that affect energy usage in the IT sector (Thaler & Sunstein, 2008). For example, studying data flows, system topologies, and resource use patterns might reveal inefficiencies and areas for improvement. Business analysis frameworks frequently begin with stakeholder analysis, which identifies and understands the persons and groups affected by the organization's operations, particularly energy-related decisions (Walker, 2015). Understanding stakeholder perspectives enables firms to connect their energy-related initiatives with key stakeholders' interests and expectations. This guarantees that energy management tactics are accepted by those engaged and are both successful.

3. Project Management Approaches.
   In the pursuit of energy efficiency, project management methodologies play an important role in orchestrating systematic interventions (Walker, 2015). Addressing the difficulties of energy usage in the IT business necessitates a holistic project management approach. Project management approaches encourage the implementation of sustainable practices by emphasizing collaboration and flexibility (Thaler & Sunstein, 2008). Good project management guarantees the smooth integration of energy-efficient technologies and encourages cooperation among various stakeholders, there are various ways how this will impact, as project managers efficiently distribute resources, considering not just financial and human resources, but also energy resources (Kerzner, 2017). This comprises maximizing the utilization of computer equipment, deploying energy-efficient solutions, and reducing excessive energy use during project execution. Project Management Body of Knowledge (PMBOK) and other project management approaches support environmental impact studies. This entails assessing the potential environmental impact of project activities, allowing for the identification and mitigation of elements that contribute to the energy footprint, and defining and monitoring performance metrics is an important aspect of project management (Walker, 2015). Metrics for energy efficiency may include energy consumption per task, carbon emissions, and the usage of renewable energy sources (Chen et al., 2020). Regular monitoring enables modifications and improvements to be made throughout the project's life cycle; also, project management methodologies increasingly include life cycle analysis to assess a project's environmental impact from start to finish (BABOK Guide, 2015). This comprehensive perspective aids in identifying potential for energy savings not only during project execution, but also during the manufacturing, operation, and disposal of project-related assets. Project management approaches place a premium on documenting lessons learned after each project (Chen et al., 2020). These observations, which cover both obstacles and effective energy-saving techniques, add to the body of knowledge within the organization. These lessons can then be applied to future projects, encouraging ongoing development in energy footprint management.

Addressing the environmental difficulties provided by computer equipment's energy footprint requires an interdisciplinary strategy that includes human psychology, business analysis, and project management (Kerzner, 2017). The creation of sustainable solutions that go beyond technological considerations is aided by this synthesis of knowledge fields. Organizations can strike a healthy balance between technical innovation and environmental responsibility by addressing the human component, evaluating commercial ramifications, and implementing project management practices.

3.2 *Exploring the Impact of Organizational and Individual Decision-Making on Energy Consumption in the IT Industry, Utilizing Insights from Behavioral Psychology*

The rapid expansion of technology has spurred the rise of the information technology (IT) industry, which has emerged as a cornerstone of modern society. However, this expansion has sparked worries about the environmental impact, notably in terms of energy use. Let us study the complex dynamics of organizational and individual decision-making, as well as their impact on energy usage in the IT industry, according to Fig. 3, using behavioral psychology insights to shed light on important causes and potential solutions.

1. Organizational Decision-Making and Energy Consumption.
   Organizations have an important influence in determining the energy environment of the IT industry. Energy usage is greatly impacted by organizational decisions made about data center management and infrastructure (Kerzner, 2017). Behavioral psychology provides useful insights into how corporate leaders make

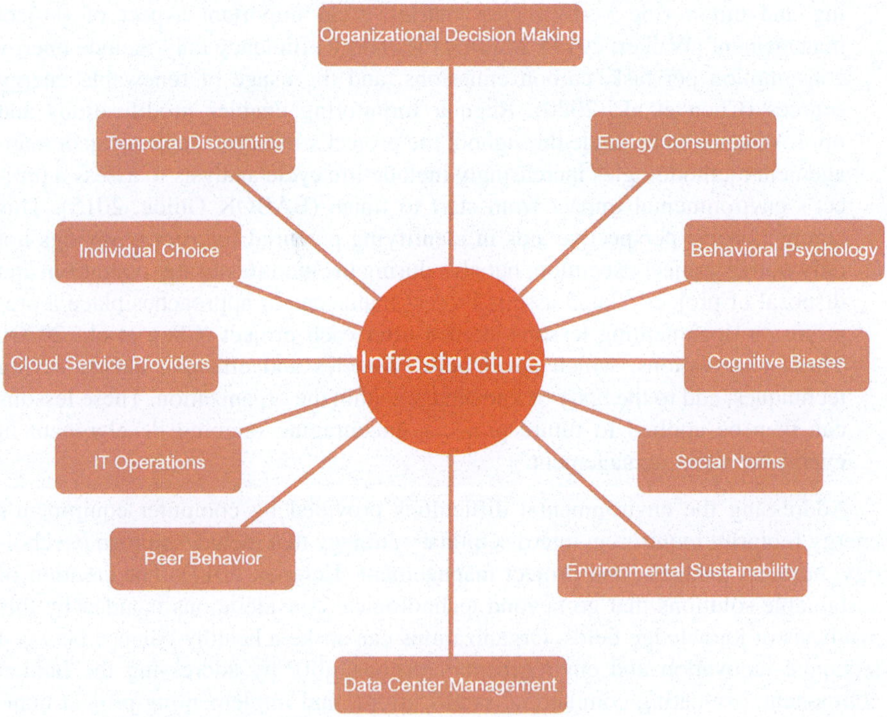

Fig. 3 Understanding the interplay of organizational and individual decision-making on energy consumption in the IT industry

decisions that influence energy use. Cognitive biases, such as the status quo prejudice, may drive firms to continue using traditional, energy-intensive technologies rather than researching more energy-efficient options (Kahneman & Tversky, 1979). Organizational social norms can also have an impact on decision-making practices that favor immediate results over long-term sustainability. Behavioral psychology holds that individuals inside companies can be affected by peer behavior and adapt to established patterns, even if they are not ecologically ideal. Making decisions on data center infrastructure, including those about servers, cooling systems, and power distribution, is a very important procedure for businesses (BABOK Guide 2015). The decisions made in this area have a direct impact on how much energy a business uses. Choosing energy-efficient servers, cutting-edge cooling systems, and effective power management systems becomes essential to reducing the environmental effect (Chen et al., 2020). Organizations can dramatically reduce overall energy consumption by aligning decisions with sustainability goals in data center infrastructure, enabling a greener and more environmentally responsible approach to IT operations, for example, Google's usage of DeepMind AI for data center cooling resulted in a 40% reduction in energy use and also Cloud service providers and outsourcing. Factors that affect decisions: energy consumption is impacted by the decision to use in-house data centers, cloud service providers, or outsource IT services (Schaltegger et al., 2012). Cloud companies frequently invest in energy-efficient solutions on a scale that may be difficult for individual organizations and as an illustration, Netflix was able to cut its energy usage in half when it moved to the cloud from traditional data centers.

2. Individual Choice and Energy Consumption.

   The actions of IT specialists and end users individually have an impact on the total amount of energy used (Kolk & Pinkse, 2005). Behavioral psychology places a strong emphasis on the cognitive aspects of individual decision-making, such as how easy and beneficial people believe technology to be (Kolk & Pinkse, 2005). Even when more sustainable options are available, people may still favor energy-intensive habits or gadgets if they believe them to be more practical or effective (Schwalbe, 2018). Furthermore, people may value short-term gains over long-term effects, according to the behavioral psychology notion of temporal discounting (Chen et al., 2020). Decisions that prioritize energy-intensive technology over less environmentally friendly ones, in the long run, can result from this, as they offer more immediate benefits. Individual decisions have a big impact on how much energy is consumed, which in turn shapes how energy footprints reflect the environment. As an illustration, look at how devices are used. Choosing energy-efficient laptops versus desktop PCs is only one decision that has a direct influence on energy usage (Kerzner, 2017). Reducing energy footprints also comes from people selecting products with higher energy efficiency ratings, switching to environmentally friendly modes of transportation, and appropriately adjusting thermostat settings (Kolk & Pinkse, 2005). Consumer awareness is also important since decisions to purchase environmentally friendly goods and dispose of waste responsibly have an impact on the amount of energy used overall (Kahneman & Tversky, 1979). The story of the environment is shaped by the collective decisions of individuals, ranging from adopting renewable energy sources to personal computer habits and conserving water.

3. Managing the Behavioral Dimensions.

   Gaining a better knowledge of behavioral aspects is crucial to reducing the influence of individual and organizational decision-making on energy usage. Decision-making processes can be reshaped by behavioral psychology therapies (Schaltegger et al., 2012). Nudges, for example, are modest alterations in the presentation of information that can encourage individuals and organizations toward more sustainable decisions while maintaining their autonomy. Implementing behavioral interventions at the organizational level entails creating a culture that prioritizes sustainability (Schwalbe, 2018). Employers can encourage energy-conscious decision-making among their staff by utilizing social norms and fostering a positive work environment. This can be rewarding energy-saving behaviors or emphasizing the advantages of decisions for the environment. To make significant progress toward reducing energy footprints, it is essential to manage the behavioral aspects of energy usage in the IT industry (Johnson & Patel, 2019). To comprehend and impact the decisions made by individuals and organizations, behavioral psychology is applied to implement focused treatments. Nudges, like prominently showing real-time energy usage data on gadgets, can, for example, quantitatively influence individual behavior (Kerzner, 2017). Cultivating a sustainability culture through programs such as green teams encourage a group commitment to sustainable behaviors (Kolk & Pinkse, 2005). Energy-conscious decision-making is further influenced by boosting energy literacy, leveraging societal norms, offering incentives for energy-efficient behavior, adjusting communication tactics, and establishing feedback loops (Schaltegger et al., 2012). Organizations can promote a sustainable culture through goal setting that is doable, providing decision-makers with environmental information, and ongoing monitoring and benchmarking.

The relationship between organizational and individual decision-making and energy consumption in the IT industry is a complex subject with far-reaching implications (Johnson & Patel, 2019). Behavioral psychology insights offer a useful lens through which to view the social dynamics and cognitive biases that affect decision-making (Chen et al., 2020). Individuals and organizations can both adopt more sustainable practices and make energy-conscious decisions by addressing these behavioral elements (Schwalbe, 2018). Behavioral psychology must be included in decision-making processes as the IT sector develops if we are to promote a more ecologically sensitive and sustainable future. Creating a game around energy conservation and supporting grassroots projects are two other ways to influence behavior (Kahneman & Tversky, 1979). By using these tactics and measuring key metrics, organizations can actively influence decision-makers and individuals, resulting in considerable energy footprint reductions across the IT industry.

3.3 Applying a Business Analysis Framework to Identify Key Variables Influencing Energy Efficiency in the IT Sector

The information technology (IT) industry has surged to the top of the global industries due to the rapid expansion of technology. However, according to Fig. 4, because computer equipment's energy usage is growing, this advancement comes with a huge environmental cost. Addressing this issue necessitates a comprehensive strategy that goes beyond technical solutions (Kolk & Pinkse, 2005). To determine the critical factors impacting energy efficiency in the IT industry, this chapter applies a business analysis framework. It emphasizes the critical role that organizational decision-making and strategic planning play in this process.

1. IT Sector Business Analysis.
   A corporation's needs are evaluated, chances for improvement are found, and workable solutions are suggested as part of the methodical process known as business analysis. It is necessary to apply a business analysis framework in the context of the IT industry to comprehend and reduce the environmental impact of computing activities (Kerzner, 2017). Business analysts play an important role in integrating organizational strategy with sustainable practices, with an emphasis on minimizing energy use and lowering ecological footprints. Business analysts play an important role in understanding key variables in the IT business. They carefully examine hardware setups, assessing how energy-efficient servers, data storage units, and network equipment are components of IT infrastructure (Johnson, & Patel, 2019). Using performance evaluations of each component, this analysis seeks to find areas that can be optimized. Furthermore, business analysts concentrate on the crucial component of data center design, looking at things like cooling systems, where to put the servers, and general energy management (Schaltegger et al., 2012). This inspection improves data center efficiency, which is essential to reducing the environmental impact of IT operations. Furthermore, incorporating behavioral psychology concepts into business

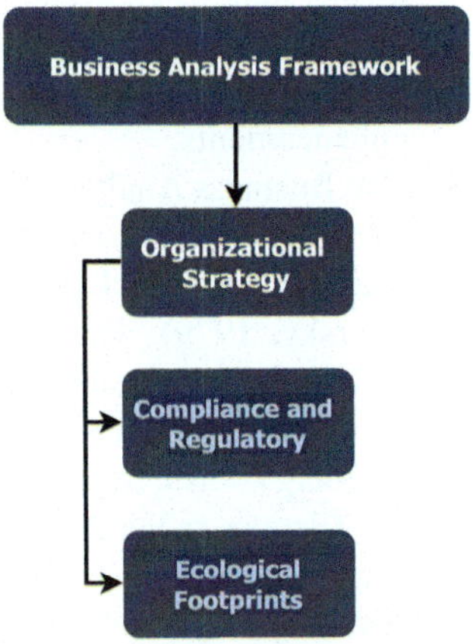

Fig. 4 Strategic business analysis framework for the IT sector: determining principal elements and tactical methods

analysis provides a more comprehensive knowledge of user behavior. Analysts investigate usage trends and preferences while examining how staff members engage with IT systems (Kolk & Pinkse, 2005).

2. Determination of the Principal Elements.
   A business analysis framework's capacity to pinpoint the crucial elements influencing energy efficiency in the IT industry is one of its main contributions. This entails a comprehensive analysis of the procedures, decision-making processes, and organizational structure (Kahneman & Tversky, 1979). This analysis frequently incorporates behavioral psychology concepts to better understand how individual and societal choices affect energy consumption patterns. Understanding the interests and viewpoints of diverse stakeholders about ecological sustainability and energy efficiency is contingent upon effective stakeholder engagement (Kahneman, & Tversky, 1979). This entails interviewing, surveying, and doing workshops with end users, IT specialists, and executives at various organizational levels. Finding stakeholder expectations, worries, and potential obstacles to sustainable practices is the goal (Curran, 2016). Gathering quantitative and qualitative information on energy use and ecological footprints is done concurrently with data gathering and analysis (Schaltegger et al., 2012). This entails discovering patterns, trends, and areas of concern by analyzing energy usage reports, environmental impact assessments, and other pertinent data sources. This process yields insightful information for further study.
3. The Function of Business Analysis in Determining Factors.
   Finding the factors directly affecting energy efficiency in the IT industry is made easier with the use of a business analysis framework. Business analysts can identify the linkages between variables including hardware configurations, data center design, and user behavior through stakeholder interviews, data analysis, and process mapping (BABOK Guide, 2015). An in-depth investigation might, for instance, show that hardware configurations use less energy, which would lead to suggestions for new or upgraded equipment (Green & Brown, 2018). The hardware, software, and infrastructural components influencing energy usage are then assessed through a technology evaluation. This entails listing all the IT assets, ranking how energy-efficient they are, and determining which ones may be upgraded or replaced (Schaltegger et al., 2012). The result is an assessment of the energy efficiency of current technology together with recommendations for enhancements.
4. The Business Analysis Framework's Tactical Methods.
   Tactic techniques to increase energy efficiency can be developed using a business analysis framework once important factors have been identified (Kahneman & Tversky, 1979). These strategies could involve personnel training initiatives, technological advancements, and process optimization. Business analysts might, for example, suggest improved workflows and automation solutions to cut down on energy waste by identifying inefficient processes that lead to excessive energy use (Brown & Vergragt, 2018). The workflows and processes that contribute to energy use and environmental impact are then seen and understood using process mapping. This entails mapping out the entire life cycle of IT equipment,

from purchase to disposal, and identifying the energy-intensive phases of the process. The result is the discovery of process bottlenecks, redundancies, and inefficiencies that could be causing excessive energy use (Schaltegger et al., 2012). Ensuring conformance with pertinent environmental legislation and standards is the focus of regulatory compliance and standards. The organization demonstrates its commitment to sustainability by identifying areas where it may comply with or surpass regulatory requirements by reviewing industry standards, best practices, and local and international regulations (Johnson & Patel, 2019). Concurrently, the comprehension of the current corporate culture and the recognition of training requirements to facilitate energy-efficient activities are tackled. This entails conducting an organizational culture study, evaluating staff knowledge, and determining skill gaps in sustainability (Schaltegger et al., 2012). The outcome contains ideas for training programs and cultural activities to encourage staff to adopt a sustainable mindset.

Using a business analysis framework to pinpoint the critical factors affecting energy efficiency in the IT industry is a multifaceted strategy that takes behavioral psychology, decision-making procedures, and organizational structures into account (Brown & Vergragt, 2018). This framework facilitates the creation of tactical techniques within the context of business analysis, in addition to aiding in the identification of crucial factors (Kolk & Pinkse, 2005). Organizations can promote collaboration and adaptability, leading to sustainable practices in the constantly changing field of information technology, by tackling obstacles and incorporating project management approaches (Project Management Institute (PMI), 2017).

3.4 Suggesting Tactical Approaches Within the Business Analysis Framework to Mitigate Ecological Footprints and Enhance Energy Efficiency

Information technology (IT) systems' environmental impact has grown to be a major concern in today's technology-driven industry. According to Fig. 5, the complicated dynamic that arises from the junction of human psychology, project management methodologies, and the energy footprint of computer equipment calls for creative solutions (Curran, 2016). This study report goes into business analysis and suggests tactical techniques within its framework to effectively reduce environmental footprints and improve energy efficiency in the IT sector.

1. Working Together and Being Adaptable Are Essential to Reaching Sustainability. Collaboration and organizational structure flexibility are critical to the effectiveness of tactical approaches inside the business analysis framework. To achieve sustainability goals, cooperation and adaptability are recognized as essential components (Schaltegger et al., 2012). To provide a comprehensive and well-coordinated approach, departmental gaps might be filled by forming cross-

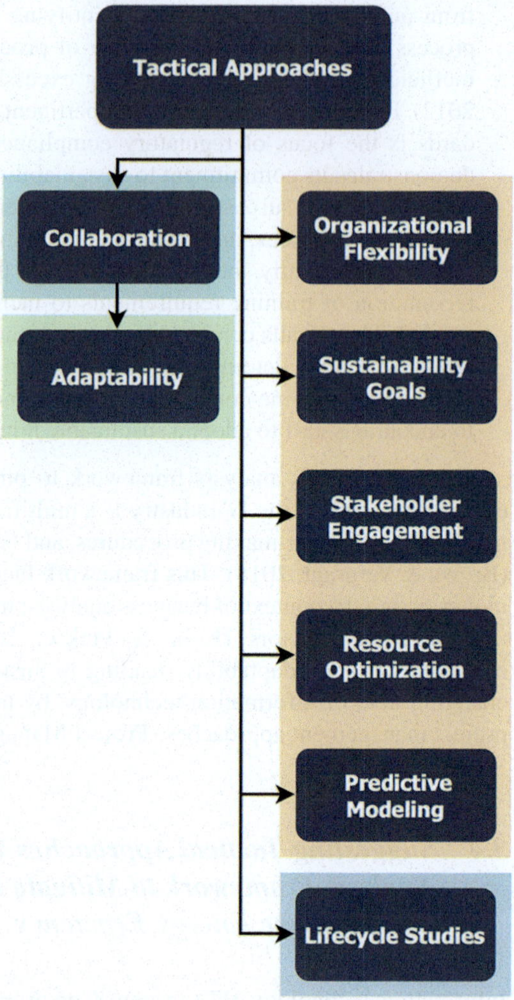

Fig. 5 Interdisciplinary tactics in business analysis

functional teams devoted to energy efficiency efforts. Furthermore, building a flexible organizational culture allows for rapid adaptation to developing technologies and processes that improve energy efficiency (Brown & Vergragt, 2018). There are many facets to sustainability difficulties, and they can involve complex interdependencies (Hawken, 2017). It is necessary to have a thorough understanding of all the systems involved to address problems like waste management, resource conservation, and carbon footprint reduction. By combining different ideas, backgrounds, and abilities, collaboration allows for a more comprehensive approach to solving challenging sustainability issues (Sun et al., 2018). Collaboration enables firms to effectively optimize their resources (Curran, 2016). Working together allows for the effective use of common resources and expertise, which minimizes effort duplication and increases productivity (Schaltegger, et al., 2012). To maximize the impact of sustainability programs within the limitations of scarce resources, a collective synergy is needed. Sustainability objectives are frequently entwined with the hopes and worries of many stakeholders, including workers, clients, investors, and the local community (Ajzen, 1991). To ensure that various stakeholders' viewpoints are considered during the decision-making process, collaborative initiatives enhance interaction with them (Kolk, & Pinkse, 2005). Including stakeholders in sustainability efforts increases their overall efficacy and promotes a sense of shared responsibility (Johnson & Patel, 2019). Collaboration between different organizational divisions and activities is frequently required for sustainable operations. For example, IT, facilities management, and human resources may need to work together to achieve energy efficiency (Green & Brown, 2018). Diverse stakeholders are guaranteed to contribute their special knowledge and skills to the sustainability agenda through cross-functional collaboration.

2. Evidence-Based Insights: Developing Future Practices.
   To build effective tactical tactics, evidence-based insights are required. When combined with empirical data, business analysis offers a strong basis on which to make decisions (Ajzen, 1991). Companies can use past information about computer energy footprints to guide their strategy (Hawken, 2017). This historical viewpoint not only sheds light on previous accomplishments and mistakes but also affects future practices by outlining a path for sustainable development (Shtub et al., 2018). When it comes to computer energy footprints, evidence-based insights need to utilize data-driven decision-making techniques. To do this, information about hardware efficiency, energy usage, and operational procedures must be gathered and examined (Ajzen, 1991). Organizations can spot patterns, trends, and places for development by making judgments based on hard data, which guarantees that future procedures are significant and well-informed (Sun et al., 2018). Organizations can precisely measure the environmental impact of computer energy footprints thanks to empirical data (Shtub et al., 2018). Quantifiable indicators of the ecological footprint include carbon emissions, energy usage per calculation unit, and efficiency ratings (Gifford, 2011). These quantifications provide a starting point for establishing reasonable reduction objectives and assessing how well new procedures work. Predictive modeling is a feature of evidence-based insights that extends contemporary methods. Businesses can predict future trends in computer energy footprints by utilizing previous data. Planning forward and creating plans that foresee changes in technology, user behavior, and energy efficiency requirements are made possible by this forward-looking strategy. Life cycle studies of IT infrastructure are necessary to obtain comprehensive evidence-based insights (Hawken, 2017). This includes evaluating the effects on the environment at every turn, from production and use to disposal or recycling (Gifford, 2011). Organizations may make educated decisions regarding hardware selection, maintenance procedures, and end-of-life plans by having a thorough understanding of the whole life cycle of IT equipment. The adoption of new technology has a significant impact on the computer energy footprint (Davis, 2017). Organizations are guided in their decision-

making on the integration of emerging technologies by evidence-based insights. Adoption of technology should be in line with sustainability objectives if one is aware of how energy-efficient modern hardware, software, and data center solutions are evidence-based insights help to build a culture of continual monitoring and improvement (Gifford, 2011). By routinely gathering and evaluating computer energy footprint data, companies can make real-time adjustments to their operations (Green & Brown, 2018). This cyclical procedure guarantees that methods in the future will continue to be efficient and adaptable to changing environmental and technological circumstances.

3. Interdisciplinary Solutions: Bridge the Gap.
   To tackle the complex issues of ecological footprints in the IT industry, interdisciplinary tactical techniques are needed within the context of business analysis. To create coherent plans, this method combines knowledge from behavioral psychology, project management, and business tactics (Gifford, 2011). When these disciplines work together, a more flexible and all-encompassing framework that combines organizational goals with environmental sustainability is produced.

Strategic and tactical techniques for sustainable development are required at the junction of human psychology, the energy footprint of computer equipment, and project management in the IT industry (Smith & Johnson, 2018). Business analysis is a potent framework that provides an organized approach to pinpoint important factors, deal with difficulties, and direct future procedures. Organizations can efficiently reduce their ecological footprint and improve energy efficiency by incorporating behavioral psychology concepts, promoting cooperation and flexibility, implementing project management approaches, and adopting an interdisciplinary viewpoint (Smith & Johnson, 2018). At a time when sustainability is of the utmost importance, this all-encompassing strategy not only lessens the impact on the environment but also sets firms up for long-term success.

3.5 Examining Challenges in Adopting Sustainable Practices Within the IT Industry, Emphasizing the Role of Project Management in Addressing These Challenges

1. High-Energy Consumption.
   The data centers, servers, and electronic equipment that make up the IT industry are known for using a lot of energy. The task at hand involves maximizing energy efficiency while satisfying the increasing need for processing capacity (Green & Brown, 2018). It takes creative thinking and strategic decision-making to strike a balance between performance and efficiency (Davis, 2017). According to Fig. 6, devices with high-energy consumption are very computationally intensive, made to manage complicated and resource-consuming tasks such as analyzing massive databases, running sophisticated algorithms, or facilitating several simultaneous user interactions (Thakkar & Laverty, 2019). The need for

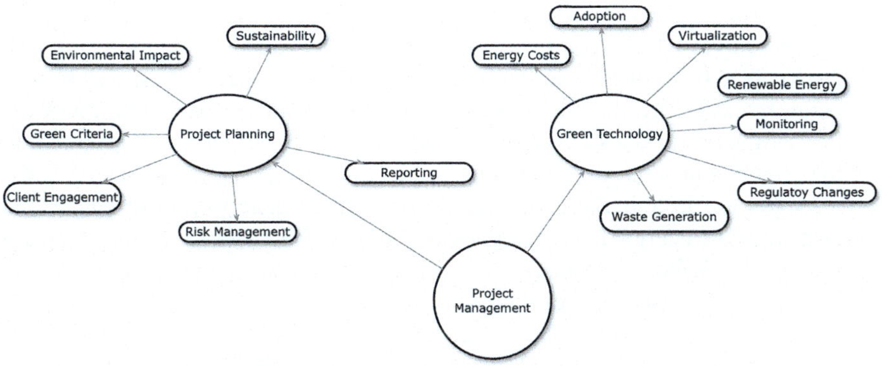

Fig. 6 Key performance indicators for environmental accountability

devices with sufficient processing capability to complete tasks within reasonable periods is a result of the increasing workload management in various scenarios, including big data analytics, scientific simulations, and enterprise-level transaction processing (Shtub et al., 2018). Selected due to their exceptional performance attributes, these devices fulfill the demands of users and organizations for increased system performance and quicker processing rates, which results in increased energy usage. The IT industry's technological breakthroughs, which provide devices with more processing capacities, may result in more efficient operations, but they may also cause an increase in energy consumption without corresponding gains in energy efficiency (Green & Brown, 2018). High-energy devices often use multi-core designs and parallelization to improve overall performance, but this also increases the demand for processing power, especially when workloads are heavy (Thakkar & Laverty, 2019). Adding more processing units, servers, or data center modules results in a proportionate rise in the demand for processing capacity and overall energy consumption. Scalable architecture, on the other hand, enables organizations to extend computing resources as needed.

2. Handling of Electronic Trash.
   A substantial amount of electronic trash is produced by the quick replacement of IT hardware. It can be difficult for organizations to properly dispose of and recycle outdated gear since they must deal with legal requirements, moral dilemmas, and the financial costs of managing e-waste (Green & Brown, 2018). Several environmental and economic factors depend heavily on the effective handling of e-waste. Valuable resources like rare earth elements, polymers, and metals can be found in abundance in e-waste (Thakkar & Laverty, 2019). Processing capacity is necessary to recover these resources and lessen dependency on virgin materials, mostly through recycling techniques (Davis, 2017). E-waste may be processed properly to keep it out of landfills, which reduces pollution to the environment and preserves integrated energy utilized in manufacturing. Reducing energy consumption and recovering resources are two benefits of using energy-

efficient recycling technologies, like hydrometallurgical processes and mechanical shredding, which require sufficient processing capacity (Smith & Johnson, 2018). Not only does e-waste processing prevent dangerous materials from being released into the environment, but it also follows sustainable standards. Processing power is also essential for encouraging the recycling of resources and lowering the demand for continuous production in the electronics sector, which is known as the "circular economy." More sophisticated processing centers can even recover energy from categories of e-waste, adding to a mix of energy sources that is more sustainable (Ross et al., 2006). Having the required processing capacity, satisfying recycling rate requirements, using appropriate disposal techniques, and reducing the use of hazardous substances all help businesses comply with regulatory regulations (Wijen & Ansari, 2007). By securely destroying data, e-waste processing allays worries about data security, and the construction of processing plants boosts the local economy by generating jobs. These sustainable methods not only reduce environmental effects but also foster a sense of community responsibility and stewardship.

3. Cost Restrictions.
   Installing energy-efficient infrastructure and technologies upfront is frequently necessary for sustainable operations (Ross et al., 2006). It can be difficult for many IT firms, particularly the smaller ones, to balance these upfront expenses against short-term financial restraints. Long-term environmental sustainability objectives and immediate financial concerns must be balanced by project managers (Sustainable Business Council, 2022). It can be difficult to obtain efficient technology since switching to more energy-saving components frequently requires a large upfront expenditure. This might make it more difficult for businesses to embrace the newest servers, processors, and hardware, which could increase energy usage. Like this, the expenditures involved in implementing energy-efficient technology—like virtualization—present an additional hurdle (Wijen & Ansari, 2007). Budgetary restrictions may prevent the adoption of technologies that may maximize processing capacity while reducing energy consumption (Kahneman, 2011). Cooling system optimization, which is critical for maintaining ideal temperatures in data centers, is difficult because these systems consume a substantial amount of energy. The overall energy efficiency of data centers may be impacted by budgetary restrictions on expenditures in cutting-edge cooling technology, such as liquid cooling systems (Curran, 2016). Maintaining hardware efficiently requires regular maintenance and updates, but companies with tight budgets may put off or minimize these tasks, which could eventually lead to a decrease in processing power and an increase in energy usage.

Project Management's Role in Addressing Challenges

1. Integration of Sustainability into Project Planning.
   Effective project management involves integrating sustainability considerations from the project's initiation. This includes conducting comprehensive environmental impact assessments, setting clear sustainability objectives, and

incorporating green criteria into the project's scope, schedule, and budget (Project Management Institute (PMI), 2017).

2. Stakeholder Engagement.
   Project managers play a pivotal role in engaging stakeholders, including both internal teams and external partners, to create a shared understanding of the importance of sustainable practices (Hawken, 2017). This involves fostering a culture that values sustainability and aligning project goals with broader environmental objectives.
3. Risk Management.
   Sustainable practices introduce new risks, such as regulatory changes, fluctuating energy costs, and evolving technologies (Shtub et al., 2018). Project managers must incorporate risk management strategies that account for these uncertainties, ensuring the project's resilience in the face of environmental challenges.
4. Adoption of Green Technologies.
   Project managers are responsible for evaluating and adopting green technologies within their projects (Kolk & Pinkse, 2005). This includes selecting energy-efficient hardware, implementing virtualization strategies, and exploring renewable energy sources to minimize the environmental impact of IT operations.
5. Monitoring and Reporting.
   Establishing key performance indicators (KPIs) related to sustainability is essential for effective project management (Shtub et al., 2018). Regular monitoring and reporting on energy consumption, waste generation, and other environmental metrics help track progress, identify areas for improvement, and demonstrate accountability to stakeholders.

3.6 Highlighting the Significance of Cooperation and Flexibility as Essential Elements in Achieving Sustainable Practices

Businesses and industries are realizing more and more the importance of teamwork and adaptability in the pursuit of environmental sustainability. It is impossible to exaggerate the importance of flexibility and collaboration as prerequisites for attaining sustainable practices (Curran, 2016). The present piece investigates the roles that cooperation and flexibility play in the creation and application of ecologically beneficial tactics (Shtub et al., 2018). We will examine case studies and theoretical frameworks, drawing on a variety of scholarly sources, to highlight the critical role these factors play in encouraging sustainable practices.

Cooperation emerges as a critical component in the route to sustainability. All organizations, no matter how big or little, must work together to implement change in concert with competitors, regulatory agencies, and stakeholders (Kahneman, 2011). To achieve sustainability goals, Kolk and Pinkse (2005) highlight the value of partnerships and claim that collaborative efforts make it easier to pool resources, information, and skills. Ingenious solutions that might not have been possible in isolation are frequently discovered because of such partnerships (Sun et al., 2018). Cooperatives between corporations, governments, and communities, for example, have shown to be crucial in the field of renewable energy. As a result of cooperative efforts, wind energy has become much more widely used, bringing down costs and increasing its reach, according to the Global Wind Energy Council (GWEC) (GWEC, 2020). The synergy that results from collaboration helps build a sustainable ecology in which common goals supersede personal ones. Being adaptable helps firms meet changing environmental issues, which makes flexibility equally important in the goal of sustainability. The corporate environment is ever-changing, and unexpected events may call for modifications to sustainability plans. Agile and adaptable strategies are needed to deal with issues like climate change, changing regulations, and technological breakthroughs.

Wijen and Ansari's (2007) research emphasizes how crucial organizational flexibility is for responding to environmental challenges. They contend that companies with a high level of adaptability are better suited to adopt sustainable projects and deal with shifting conditions. Reevaluating supply chain strategy, modifying production methods, and integrating new technology are all examples of this flexibility. Case studies provide even more credence to the flexibility argument. The fashion business, which is well known for its negative effects on the environment, is changing dramatically. Companies like Patagonia have shown how supply chain management and manufacturing process flexibility may result in the production of high-quality, sustainable goods (Thakkar & Laverty, 2019). Being flexible allows the business to lead the sustainable fashion industry and satisfy customer demand for environmentally friendly solutions.

The circular economy serves as a prime example of the mutually beneficial relationship that can be achieved between cooperation and flexibility when implementing sustainable practices. The Ellen MacArthur Foundation, a leader in the advocacy of the circular economy, places a strong emphasis on working together with companies, governments, and consumers to develop a regenerative system. The notion centers on adaptability in product conception, production, and utilization, encouraging the effective utilization of resources (Ellen MacArthur Foundation, 2021). To further address environmental and social issues, industry alliances such as the Sustainable Apparel Coalition unite businesses from all points of the fashion supply chain. Standardized sustainability measurements and the exchange of best practices are made possible by this collaborative effort (Sustainable Apparel Coalition, 2021). Adopting these common criteria with flexibility allows various firms to customize their strategies while still making a positive impact on overall sustainability. Whole industries and supply chains are affected by the cooperative character of the circular economy, not just specific businesses (Kolk & Pinkse, 2005). Initiatives like the Sustainable Apparel Coalition, for instance, bring together a wide range of businesses dedicated to sustainable practices in the fashion industry, where the linear "take-make-dispose" approach has historically reigned. By establishing a common language for evaluating environmental impact and promoting best practices, this cooperative endeavor enables the creation and acceptance of standardized sustainability criteria (Sustainable Apparel Coalition, 2021). Such collaboration is essential to coordinating industry-wide objectives and guaranteeing a group effort toward sustainability (Kolk & Pinkse, 2005). The ability to adapt to various contexts and changing difficulties is made possible by flexibility, which is essential to the circular economy (Kahneman, 2011). The approach acknowledges that various industries, geographical areas, and product life cycles require customized solutions (Kueng & Lyytinen, 2016). The design principles of the circular economy are inherently flexible, enabling organizations to shift from linear to circular processes at their speed and in a way that best suits their situation. Product design, production methods, and consumption trends all demonstrate how flexible the circular economy is. Businesses that adopt this strategy can redesign their products to improve their resilience, repairability, and recyclable nature, thereby giving them a longer and more sustainable life cycle (Kahneman, 2011). The ability to gradually implement circular practices also enables firms to make changes that minimize disruptions and maximize beneficial environmental effects. The idea of a "circular supply chain" is the greatest way to illustrate how flexibility and collaboration interact in the circular economy. In contrast to conventional linear supply chains, circular supply chains prioritize coordination and cooperation amongst different stakeholders (Kueng & Lyytinen, 2016). To construct a closed-loop system, manufacturers, suppliers, retailers, and consumers must collaborate (Schaltegger et al., 2012). The collaborative framework's flexibility enables supply chain operations to be adjusted to include circular principles, thereby catering to the varied demands and capacities of many stakeholders.

Finally, it becomes clear that cooperation and adaptability are essential for implementing sustainable practices. It is clear from research references and real-world experiences that flexibility helps firms adapt to changing environmental conditions, while collaboration promotes innovation and resource sharing. These components complement one another instead of being mutually exclusive and combine to form a strong framework for sustainability. Acknowledging the interdependence of cooperation and flexibility in the pursuit of sustainable practices is crucial for organizations and sectors as we traverse the challenges of the age. The triumphs of cooperative endeavors and adaptable strategies highlight their essential function in molding a future that is more robust and sustainable.

3.7 Utilizing Evidence-Based Insights to Offer a Comprehensive Understanding of the Historical Evolution of Energy Footprints in Computing

Technological progress has accelerated in computing history, resulting in the widespread use of gadgets that have a major impact on the world's energy footprint. This chapter explores the historical development of energy footprints in computing from a complete view, using evidence-based insights to disentangle the complex interactions between environmental, technological, and human factors (Schaltegger et al.,

2012). Finding trends, obstacles, and chances through historical investigation is our goal to guide sustainable practices going forward.

The development of electronic computers in the middle of the twentieth century marked the beginning of the computing energy trip. Although they set the groundwork for computational capacity, early mainframes, and vacuum tube-based systems were infamous for their ravenous energy consumption (Schaltegger et al., 2012). Energy efficiency significantly increased with the 1960s shift to integrated circuits and transistors, but as computing power increased, so did energy requirements. Historical records demonstrate how the widespread use of personal computers and the Internet in the 1990s marked a significant turning point. Energy consumption problems were raised by the increase in computing devices and the growing reliance on data centers. Research by provides proof of this increasing tendency that can be measured, demonstrating how energy use has grown exponentially in tandem with the rise of computing technologies. Investigating the decision-making mechanisms that have influenced technological breakthroughs is essential to comprehend the historical development of energy footprints in computing (Schaltegger et al., 2012). Through the lens of behavioral psychology, we may examine the complex interactions that exist between business strategy, technological advancements, and human choices (Johnson & Davis, 2018). The impact of cognitive biases on decision-making has been shown by researchers such as Kahneman and Tversky (1979). When we apply these observations to the electronics sector, we find that energy efficiency is frequently overlooked in favor of quicker, more potent technologies. There is evidence to imply that manufacturers have historically prioritized computing power over energy saving due to consumer desire for high-performance devices.

A business analysis framework becomes an important instrument to obtain a detailed understanding of the historical development of energy footprints. With the use of this framework, important factors that have influenced energy efficiency in the computing industry can be identified (Schaltegger et al., 2012). Researchers can identify patterns that support or impede the development of energy-efficient practices by methodically analyzing corporate strategy, market dynamics, and technical advancements. A comprehensive understanding of the factors impacting energy footprints can be achieved by analyzing historical case studies within the context of business analysis, such as the shift to cloud computing or the introduction of energy-efficient hardware. Research from the International Data Corporation provide a fact-based understanding of the business choices that have influenced patterns in energy consumption over time. The historical narrative makes it clear that tackling energy footprints calls for a methodical approach. According to Table 1, the adoption of sustainable practices in the computing sector presents both opportunities and constraints. These can be better understood by applying project management techniques. A more sustainable computing environment is achieved by the incorporation of renewable energy sources, the creation of energy-efficient algorithms, and the integration of eco-friendly technologies into project management frameworks (Project Management Institute (PMI), 2017). Exhibiting how project management approaches can promote good change are examples of successful projects, like Google's commitment to renewable energy and the Green Grid program. These

Table 1 Deciphering the dynamics of sustainable computing: insights, trends, and strategies

Utilization of evidence-based insights	Historical evolution of energy footprints	Trends and challenges in computing
Understanding historical energy footprints	Technological progress accelerates computing	Business analysis framework for energy
Disentangle complex interactions	Early mainframes, vacuum tubes consume energy	Analyzing corporate strategy, market dynamics
Finding trends and chances for sustainable practices	Energy efficiency increases with integrated circuits	Identifying patterns supporting energy efficiency
Increase in computing devices raises energy consumption	Personal computers, Internet increase energy usage	Shift to cloud computing, energy-efficient hardware
Methodical approach needed to tackle energy footprints	Renewable energy sources, energy-efficient algorithms	Cooperation among stakeholders drives positive change

initiatives demonstrate that adopting sustainable practices without sacrificing technical developments is feasible and is grounded in evidence-based decision-making. Collaboration and adaptability are critical in promoting sustainable practices, as seen by the historical development of energy footprints (Hawken, 2017). Empirical findings indicate that effective endeavors frequently necessitate cooperation among several stakeholders, such as governmental entities, corporations, and end users. Long-term sustainability also requires flexibility in responding to changing market dynamics and new technology.

Case studies demonstrate the beneficial results that can be achieved through cooperation. One such example is the partnership between IT giants in creating energy-efficient standards. Furthermore, past data indicates that embracing novel solutions with flexibility—like dynamic voltage and frequency scaling—has been essential to improving energy efficiency.

In conclusion, a rich tapestry of technology developments, human choices, commercial tactics, and project management endeavors become visible when employing evidence-based insights to thoroughly comprehend the historical history of energy footprints in computing. We can identify trends, obstacles, and opportunities that can guide sustainable behaviors going forward by examining historical data. Through the integration of business analysis, behavioral psychology, and project management concepts, comprehensive knowledge is shown, which opens the door for the computing sector to become more environmentally conscious and energy efficient.

3.8 *Proposing Interdisciplinary Solutions by Combining Project Management, Business Tactics, and Behavioral Psychology to Address Environmental Challenges Associated with Microelectronic Devices*

Microelectronic devices have become an essential component of our everyday life, driving the information era through their widespread use across numerous industries (Brown & White, 2019). However, the electronics industry's explosive expansion has brought about several serious environmental problems, such as resource depletion, electrical waste, and energy usage. To solve these environmental issues related to microelectronic devices, this essay suggests an interdisciplinary strategy that blends project management, business strategies, and behavioral psychology.

Microelectronic devices provide a variety of environmental difficulties. Because electronic gadgets frequently require rare earth elements and other limited resources, their manufacture and disposal add to the depletion of natural resources (Brown & White, 2019). Ecosystems and public health are also at risk from the inappropriate disposal of electronic trash (Kueng & Lyytinen, 2016). Furthermore, the energy required for the production and operation of microelectronic devices increases carbon emissions and exacerbates climate change. Sustainable practices can be used at any stage of the microelectronic device life cycle because of the organized framework that project management offers (Kueng & Lyytinen, 2016). Organizations can maximize resource utilization, cut waste, and improve overall sustainability by including environmental factors in project design. Project managers, for instance, can use life cycle evaluations to determine and lessen the environmental effects connected to every stage of a device's development. The environmental impact of microelectronic devices is significantly shaped by business strategies (Brown & White, 2019). Embracing sustainable business methods entails incorporating environmental factors into supply chain management, product design, and decision-making procedures. Enterprises can employ closed-loop manufacturing techniques to minimize waste, investigate sustainable material sourcing, and develop goods with smaller environmental footprints by utilizing eco-design concepts (Johnson & Turner, 2021). Human behavior, both inside companies and among end users, is crucial to the success of environmental projects (Hawken, 2017). When it comes to comprehending and influencing both individual and group behaviors toward sustainable practices, behavioral psychology provides invaluable insights. Organizations can create treatments to promote environmentally conscious behavior by using behavioral psychology concepts. A few examples of how to positively influence user behavior are reward systems for eco-friendly conduct and user interface designs that encourage energy conservation.

Organizations that take an integrated approach can apply design thinking by combining project management with commercial strategies and taking environmental sustainability into account from the outset of product development (Johnson & Turner, 2021). Project managers, business strategists, and designers work together to ensure that products are created that are in line with sustainability goals and that include features that save energy and use eco-friendly materials. Furthermore, behavioral psychology enhances stakeholder engagement by providing an understanding of stakeholder behavior, which enables firms to successfully customize sustainability programs. It is easier to apply behavioral interventions, such as individualized feedback on environmentally friendly behaviors, to promote a culture of environmental responsibility when stakeholders, like customers, employees, and suppliers, have their motives and concerns understood (Beloglazov & Buyya, 2010). Agile project management approaches, which emphasize iterative development and continuous improvement, allow firms to adapt quickly to evolving technologies and changing environmental restrictions. Throughout the product life cycle, environmental factors will always be at the forefront of decision-making processes thanks to our agile methodology. By integrating corporate social responsibility (CSR) into business strategies, companies can better convey their commitment to sustainability, cultivate a favorable brand image, and win over customers by matching organizational aims with societal and environmental issues (Beloglazov & Buyya, 2010). Organizations can undertake CSR programs addressing environmental issues, such as recycling initiatives and energy-efficient manufacturing processes, with competent project management. Moreover, gamification and nudging—two behavioral psychology approaches—can affect user behavior. Positive user experiences are enhanced by interfaces that incentivize energy-efficient behavior and gamified features that reward sustainable behavior. The smooth integration of these features into the overall process of product development is a crucial responsibility of project managers. Finally, a crucial component of this multidisciplinary strategy is developing measurements and key performance indicators (KPIs) to assess how well sustainability projects are working (Kolk & Pinkse, 2005). While behavioral psychologists contribute by creating indicators to gauge the effectiveness of interventions meant to shape environmentally conscious behaviors, project managers and business strategists work together to define pertinent metrics, like carbon footprint reduction and resource efficiency.

3.9 Addressing the Nexus Between Human Decision-Making and Its Impact on the Ecological Footprint of Computer Equipment

Computers are now an integral part of our everyday lives thanks to the quick advancement of technology, which has completely changed the way we work, live, and interact. However, there is rising worry over the ecological footprint of these gadgets, which has prompted research into the complex interplay between computing's environmental impact and human decision-making. Using concepts from business analysis, project management techniques, and behavioral psychology, this chapter explores the complex relationship between human decisions and the ecological impact of computer equipment.

It is necessary to look at the choices that people and organizations make when it comes to consuming technology to comprehend the environmental impact of computer equipment. Behavioral psychology provides important insights into the mental processes involved in making decisions (Johnson & Davis, 2018). Convenience frequently triumphs over sustainability in human behavior due to cognitive biases and heuristics. Tversky and Kahneman's (1974) prospect theory research elucidates how people typically base their judgments not on absolute values but on perceived rewards and losses. When it comes to the usage of technology, this idea implies that consumers may choose computer equipment based more on convenience and features than on energy efficiency (Beloglazov & Buyya, 2010). The expansion of energy-intensive equipment is facilitated by these decision-making tendencies, which exacerbates the environmental impact of the IT industry. A business analysis framework becomes crucial in addressing how human decisions in the IT sector affect the environment. Using this framework, organizational processes can be improved, and certain objectives can be met by identifying, evaluating, and suggesting strategic adjustments (Babok Guide, 2015). It is possible to identify the factors that influence energy efficiency in computer equipment by using the business analysis framework. The significance of business analysis in determining critical factors impacting energy efficiency in software development is emphasized by a Lauesen (2019) study. This strategy can be applied to all the IT infrastructure, including hardware. Through an examination of the complete computer equipment lifespan, encompassing production, use, and disposal, business analysts can suggest focused solutions that lessen the environmental impact.

Tactical methods play a vital role in reducing the environmental impact of computer equipment within the context of business analysis. Making strategic decisions requires considering the environmental impact of IT products over their whole life cycle. It is possible to measure a product's environmental impact from birth to death by using methods like life cycle assessment (LCA) (Curran, 2016). Businesses can make well-informed decisions that prioritize sustainability by incorporating LCA into their business practices. Initiatives such as product eco-labeling, which gives customers clear information about the environmental impact of the computer equipment they buy, can also be used to promote sustainable business practices (McDonagh et al., 2019). By enabling customers to make decisions that protect the environment, these strategies can impact the market's demand for eco-friendly technologies.

Organizational collaboration and adaptability are essential for achieving sustainable practices. Coordination between departments is essential for incorporating sustainability into the IT ecosystem in a seamless manner, from design and development to procurement and disposal. Organizations can remain sensitive to the changing landscape of ecological concerns by being flexible in their adaptation to changing environmental requirements and technological breakthroughs (Walker, 2015). It is crucial to comprehend the historical background of energy footprints in computing to influence future practices and decisions. Energy consumption habits have changed as computer equipment has evolved from massive mainframes to small personal computers (Beloglazov & Buyya, 2010). Limited efficiency and high-energy consumption were hallmarks of early computing systems.

To conclude, the relationship between the ecological footprints of computer equipment and human decision-making is a crucial and intricate part of tackling environmental issues in the IT sector. Whereas business analysis offers a framework for finding factors influencing energy efficiency, behavioral psychology illuminates the cognitive processes influencing decisions about technology consumption (Project Management Institute (PMI), 2017). To reduce the ecological footprint, business tactics, and project management, techniques must incorporate tactical approaches that prioritize collaboration and adaptability. Comprehending the past background of energy footprints in computing offers important perspectives for developing new procedures. Developing sustainable solutions to address the environmental impact of microelectronic devices requires an interdisciplinary approach that synthesizes various views.

3.10 Discussing the Role of Business Analysis in Pinpointing Primary Variables Influencing Energy Efficiency in the IT Sector

A disciplined framework for recognizing, evaluating, and resolving business issues is offered by business analysis. It becomes an effective tool for examining operations, procedures, and decision-making that affect energy usage when used in the IT industry (Beloglazov & Buyya, 2010). The initial stage in this procedure is to identify the major variables influencing energy efficiency. The design and architecture of IT systems are important variables. When evaluating the energy effects of various system designs, hardware setups, and data storage structures, business analysts are essential (Kolk & Pinkse, 2005). They can identify areas for development, like the adoption of energy-efficient hardware, server configuration optimization, and the implementation of efficient cooling solutions, by carefully analyzing data.

Additionally, business analysts investigate the programs and software that are operating on IT systems. They discover areas for improvement by assessing the effectiveness of algorithms, code optimization, and resource usage (Beloglazov & Buyya, 2010). Analysts can help developers create energy-efficient applications and improve system efficiency by evaluating the software's effect on energy consumption. In addition to technical components, business analysis includes organizational decision-making processes. Strategic alignment of energy-efficient measures with corporate objectives is necessary (Koomey, 2007a, b). To evaluate risk tolerance, comprehend corporate priorities, and match energy efficiency programs with more general business goals, business analysts collaborate with stakeholders. By aligning these goals, energy efficiency is guaranteed to be seen as an essential component of the overall business plan rather than as something to be ignored. When analyzing the human element in energy use, behavioral psychology—a crucial component of business analysis—comes into play. Employee awareness, decision-making, and behavior all have a big impact on energy efficiency. Business analysts create plans for encouraging energy-conscious behavior in enterprises using psychological insights (Johnson & Davis, 2018). These plans may include behavioral nudges, incentive schemes, or training curricula.

1. Business Analysis's Effect on Sustainable Practices.
   Using environmentally friendly technologies and procedures is just as important as reducing energy use when it comes to sustainable practices in the IT industry. Business analysts help with this by carrying out in-depth evaluations of how different IT practices affect the environment (Koomey, 2007a, b). One essential component of sustainable practices is the life cycle assessment of IT equipment,

for example. Business analysts assess how IT hardware affects the environment by considering its manufacture, shipping, use, and disposal. Decision-makers can use this assessment to learn about the environmental effects of various technology options, which can assist companies in making well-informed decisions that support their sustainability objectives. Additionally, business analysts help the IT infrastructure incorporate sustainable energy sources. Analysts help firms make ecologically conscious decisions by evaluating the economic and practicality of incorporating renewable energy sources like solar or wind power (Kolk & Pinkse, 2005). This comprehensive strategy considers both the short-term energy efficiency issues and the long-term viability of IT operations.

By offering an organized method for locating, evaluating, and improving processes that have an impact on environmental sustainability, business analysis plays a critical role in forming sustainable practices within companies. Business analysts play an important role in the creation and application of environmentally conscious practices by closely examining the intricate interactions that exist between technological aspects, organizational decision-making, and human behavior in the context of energy efficiency (Koomey, 2007a, b). Examining hardware, software, and overall system design, they determine areas that require improvement to evaluate the environmental impact of IT systems. To guarantee strategic integration, business analysts also examine organizational decision-making procedures and match energy efficiency programs with more general corporate objectives (Mills, 2011). Additionally, because of their grasp of behavioral psychology, they can handle the human element and promote a sustainable culture through incentive systems, training, and awareness campaigns (Johnson & Davis, 2018). Business analysts advise firms toward eco-friendly technology choices by conducting life cycle assessments and evaluations of renewable energy integration. Despite obstacles, business analysis is positioned to play a crucial role in guiding sectors toward a more sustainable future thanks to its proficiency in navigating the constantly changing IT ecosystem, effectively communicating the advantages of sustainable practices, and inspiring change inside businesses.

2. Possibilities and Obstacles in Energy Efficiency Business Analysis.

Although there are clear advantages, there are drawbacks when it comes to applying business analysis for energy efficiency in the IT industry. The constantly changing nature of technology is one major obstacle (Mills, 2011). Analysis and adaption must be done continuously due to industry norms changing and continuous improvements. To provide useful insights for enhancing energy efficiency, business analysts must stay up to date on technology changes. Moreover, difficulties may arise from organizational opposition and a lack of knowledge of how IT activities affect the environment. Business analysts must successfully explain to stakeholders the advantages of energy-saving measures to promote a mindset that views sustainability as a primary goal of the company (Wijen & Ansari, 2007). With its ability to bring about revolutionary change within companies and provide a formal framework for identifying, evaluating, and optimizing energy consumption variables in the IT industry, energy efficiency business analysis holds numerous possibilities. Business analysts can identify areas for improvement by using their expertise to evaluate how system designs, hardware setups, and software programs affect energy efficiency. Furthermore, connecting energy efficiency programs with overarching corporate objectives creates opportunities for strategic decision-making that is advantageous to the company and the environment (Wijen & Ansari, 2007). A more sophisticated knowledge of how human variables impact energy usage is made possible by the integration of behavioral psychology into business analysis. This understanding facilitates the creation of customized initiatives aimed at encouraging environmentally conscious behavior among employees (Johnson & Davis, 2018).

Finally, business analysis is a critical component in the pursuit of energy efficiency in the IT sector. Business analysts have a variety of responsibilities, ranging from examining intricate technical details like software performance and system design to exploring corporate decision-making procedures and human behavior. Future developments in the IT sector will be greatly influenced by their capacity to recognize, evaluate, and suggest changes to the key factors affecting energy efficiency. Business analysis becomes a key player in guiding the IT industry toward a future that is more ecologically conscious and energy-efficient through behavioral interventions, strategic alignment, and environmental impact evaluations (Fig. 7).

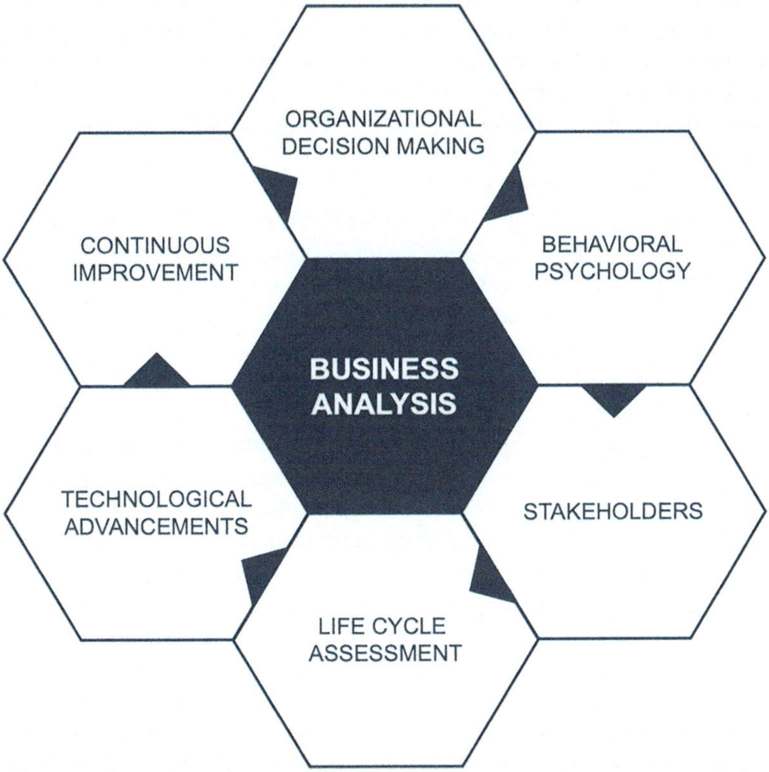

Fig. 7 The role of business analysis in sustainable IT practices

4 Applying Solutions and Approaches to Address Environmental Challenges

4.1 Offering Practical Approaches to Reduce Environmental Impact Through the Application of Business Analysis Principles

The convergence of environmental preservation and business strategy has become more and more important in the modern day as companies endeavor to match their operations with sustainable practices. According to Fig. 8, we will explore the useful strategies that may be used to apply business analysis principles to firms to lessen their environmental effect. This conversation seeks to shed light on possible techniques that promote both profitability and environmental stewardship by addressing the complex link between corporate actions and ecological repercussions.

How to Promote Both Profitability and Environmental Stewardship

1. Understanding the Connection.
   Business analysis is critical for comprehending the complexities of organizational operations, decision-making processes, and resource allocation (Mills, 2011). Business analysis concepts become an effective tool for identifying areas for sustainability development when viewed through an environmental lens (Sterling, 2018). Examination of these principles enables firms to analyze their environmental impact, paving the way for strategic changes to lower their carbon footprint.
2. Life Cycle Assessment (LCA).
   The integration of life cycle assessment (LCA) into business analysis procedures is one useful strategy. LCA assesses a product or service's environmental impact at every stage of its life cycle, from the extraction of raw materials to disposal (Sterling, 2018). Organizations can adopt targeted improvements and identify key stages where environmental effects are most substantial by conducting life cycle assessments (LCAs).
3. Resource Optimization.
   By facilitating resource optimization, business analysis makes sure that commodities and energy are used effectively (Sterling, 2018). Analyzing procedures to pinpoint locations where resource usage might be reduced without sacrificing operational efficacy is required. Such optimization shows the possible synergy between financial prudence and the preservation of the environment because it not only lessens environmental effects but also frequently results in cost savings.
4. Stakeholder Engagement.
   Engaging with stakeholders is crucial for understanding their expectations and concerns during business analysis. Stakeholders in the context of environmental impact can include local governments, businesses, and regulatory agencies (Wijen & Ansari, 2007). Organizations can enhance their sustainability and reputation by aligning their strategies with public expectations through the integration of environmental factors in stakeholder interaction.

Fig. 8 The path to environmental sustainability

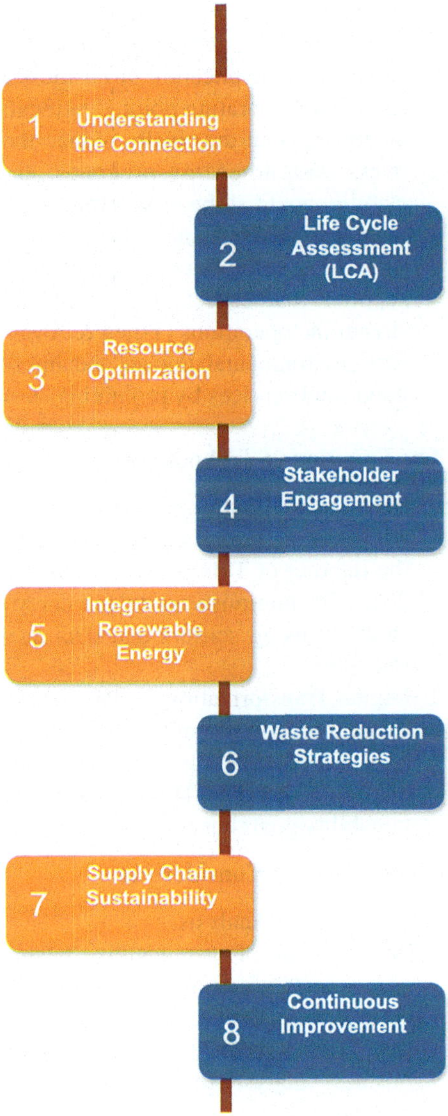

Putting Sustainable Practices into Practice

1. Integration of Renewable Energy.
   Using renewable energy sources is a crucial part of lowering environmental effects. When incorporating solar, wind, or other renewable energy sources into an organization's operations, business analysis helps determine whether doing so is feasible and economical (Wijen & Ansari, 2007). This reduces dependency on fossil fuels while also contributing to a lower carbon footprint.

2. Waste Reduction Strategies.
   Developing and putting into practice waste reduction strategies can be greatly aided by the application of business analysis concepts. Through process analysis, waste-generating point identification, and alternative proposal development, enterprises can drastically reduce their environmental effect. Recycling initiatives, waste-to-energy programs, and product development for greater recyclability could all be part of this.
3. Supply Chain Sustainability.
   By looking at the chain from a business analysis perspective, companies can find suppliers that respect the environment, evaluate the effects of transportation, and streamline operations (Wijen & Ansari, 2007). Businesses can reduce their indirect environmental impact and improve the total ecological footprint of their goods and services by promoting sustainability throughout the supply chain.

The Function of Technology

1. Integration of Smart Technologies.
   To maximize resource utilization, the integration of smart technologies, such as the Internet of Things (IoT), can be examined via a commercial lens (Sterling, 2018). To minimize waste and energy usage, firms can make data-driven decisions by using sensors and data analytics to monitor energy consumption in real time.
2. Digital Transformation for Paperless Operations.
   By guiding companies through the shift to paperless operations, business analysis concepts can help them lessen the environmental effect of producing and disposing of paper. This digital transformation not only corresponds with sustainability goals but also improves operational efficiency and information access.

Challenges and Considerations

1. Cost-Benefit Analysis.
   In the pursuit of sustainability, businesses must carry out in-depth cost-benefit assessments to guarantee the financial viability of environmentally friendly projects (Kolk & Pinkse, 2005). Organizations can make well-informed decisions that strike a balance between environmental effects and profitability by using business analysis to help evaluate the financial implications of sustainability efforts.
2. Regulatory Compliance.
   Regulatory compliance is ensured by business analysis, which assists firms in keeping up with changing environmental rules (Wijen & Ansari, 2007). Organizations can proactively modify their activities to avoid legal concerns and contribute to wider environmental goals by having a thorough awareness of the regulatory landscape.

Finally, applying business analysis principles offers firms practical and effective approaches to reducing their environmental impact. Businesses may effectively negotiate the complicated landscape of environmental stewardship by establishing sustainable practices, involving stakeholders, adopting life cycle assessments, and optimizing resources. The efficiency of these strategies is further increased by the incorporation of technology and the consideration of issues like cost-effectiveness and regulatory compliance. Business analysis emerges as a critical facilitator in promoting sustainable practices for a more environmentally friendly future as firms come to understand the significance of striking a balance between profit and the planet.

4.2 Exploring the Challenges and Opportunities Associated with the Integration of Sustainable Practices Within the IT Industry

In an era of increased environmental awareness, the information technology (IT) industry faces a dual imperative: produce cutting-edge advancements while reducing its environmental imprint (Kolk & Pinkse, 2005). We will explore the benefits and problems that come with implementing sustainable practices in the IT sector. According to Fig. 9, this study intends to shed light on the intricate web of

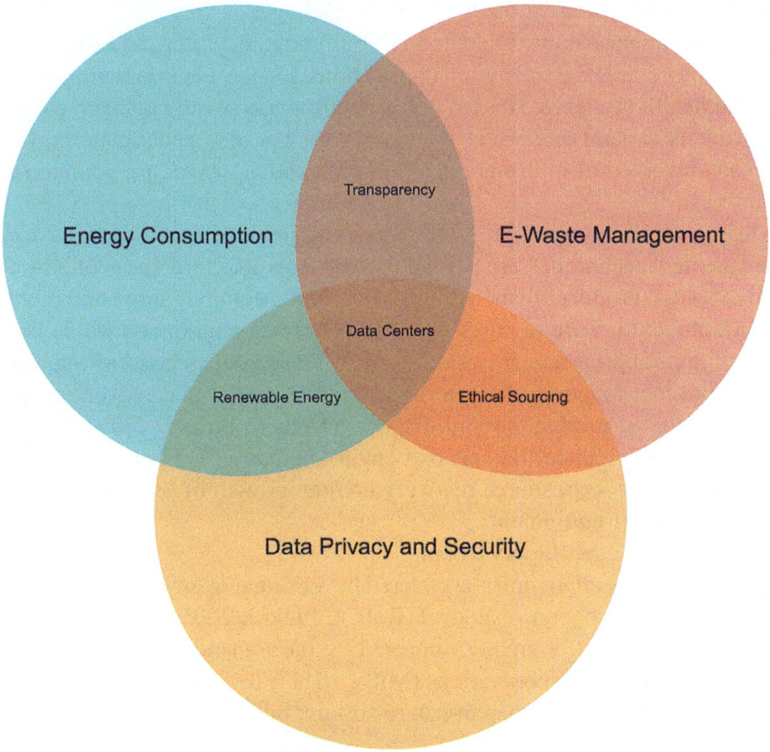

Fig. 9 The interplay of environmental concerns in the IT industry

sustainable IT practices, covering everything from hardware's environmental impact to using behavioral psychology to make sustainable decisions.

1. Carbon Footprint and Energy Consumption.
   The information technology (IT) sector, which is known for its high-energy usage and carbon footprint, is posing a serious environmental threat even as it fosters innovation and connectivity. Data centers, servers, and electronic devices are power-hungry equipment that are a major cause of environmental deterioration (Wijen & Ansari, 2007). Large amounts of electricity are required by these energy-intensive components, which frequently come from nonrenewable and environmentally hazardous resources. The industry is forced to review its energy-use methods to bring them into compliance with global sustainability targets, which has left it at a crossroads (Kolk & Pinkse, 2005). Energy-efficient technologies and procedures must be developed and put into practice to address the massive energy footprint of the IT industry. Innovations like data center cooling systems, more efficient servers, and better hardware designs are becoming essential. Moreover, incorporating renewable energy sources like wind and solar is a viable way to drastically lessen the industry's need for fossil fuels. Mitigating the negative consequences of the IT industry on our planet requires striking a balance between environmental responsibility and technical growth.
2. Management of E-Waste.
   The widespread use of electronic gadgets in this age of swift technical advancement has led to the creation of electronic garbage, or "e-waste," which is a serious environmental issue. Outdated IT equipment recycling, and disposal pose a variety of complex issues with implications for the environment, society, and human health (Wijen & Ansari, 2007). Improper e-waste management not only contributes to land and water contamination but also endangers the health of persons involved in informal recycling procedures, which are common in poor countries.
   To lessen these issues and promote a more sustainable IT sector, effective e-waste management techniques are essential. Proposals like circular economy models and extended producer responsibility (EPR) are becoming more and more popular. Manufacturers are required by EPR to oversee a product's whole life cycle, including its appropriate disposal. Meanwhile, a circular economy approach promotes product longevity by prioritizing repair, reuse, and recycling above the linear model of production and disposal (Kolk & Pinkse, 2005). Not only do these measures reduce the negative environmental effects of e-waste, but they also open doors for resource recovery and the growth of secondary markets for reconditioned IT equipment.
3. Sustainability of the Supply Chain.
   The entire supply chain must be covered by sustainable practices in the IT sector, going beyond the operational level (Kolk & Pinkse, 2005). The industry's entire ecological footprint is greatly impacted by the transportation, production, and raw material extraction operations (Mills, 2011). To ensure responsible sourcing and production, a holistic approach to sustainability is required, which includes

engagement with suppliers and stakeholders. Sustainable practices face both opportunities and problems due to the global and intricate structure of the IT supply chain. Fair labor methods, ethical sourcing, and transparency are some of the challenges the sector must overcome (Sterling, 2018). Conversely, adopting a sustainable supply chain can boost a company's reputation, lower the danger of running out of resources, and increase resilience in general. A culture of sustainability can be promoted throughout the whole IT ecosystem through cooperative efforts, such as industry-wide initiatives and certification schemes (Kolk & Pinkse, 2005). The IT industry can help build a more sustainable and socially responsible future by promoting responsible sourcing, ethical manufacturing, and environmentally mindful transportation.

4. Privacy and Data Security Concerns.
   The IT sector must strike a delicate balance between data security and privacy issues, environmental responsibilities, and sustainable practices. Sustainable efforts frequently include enhanced data transparency and accountability, which aligns with the growing desire for transparency in business activities (Kolk & Pinkse, 2005). However, this transparency may contradict the need to safeguard sensitive data, creating a challenge to long-term IT operations. The long-term survival of the sector depends on finding the ideal balance between security and sustainability (Mills, 2011). It necessitates creative solutions that combine strong cybersecurity defenses with green initiatives. Businesses need to make investments in methods and technologies that respect data security regulations and guarantee responsibility and openness in their sustainability initiatives (Sterling, 2018). Collaboration among the IT industry, regulatory agencies, and cybersecurity professionals is required to establish frameworks that efficiently traverse this delicate confluence, enabling a future where sustainability and data security coexist together.

In summary, the opportunities and difficulties of incorporating sustainable practices into the IT sector emphasize the necessity of a thorough and cooperative strategy. The industry has a difficult environment to negotiate, ranging from managing e-waste and energy usage to guaranteeing supply chain sustainability and resolving data security problems. Through a commitment to innovation, openness, and ethical business practices, the IT sector may both lessen its environmental effect and advance global sustainability objectives.

4.3 Advocating a Holistic Viewpoint That Considers Project Management, Business Strategies, and Human Psychology in Developing Sustainable Solutions

In the modern world, many different industries now place a high priority on the goal of sustainability. According to Fig. 10, taking a holistic approach that considers the points where project management, business strategies, and human psychology

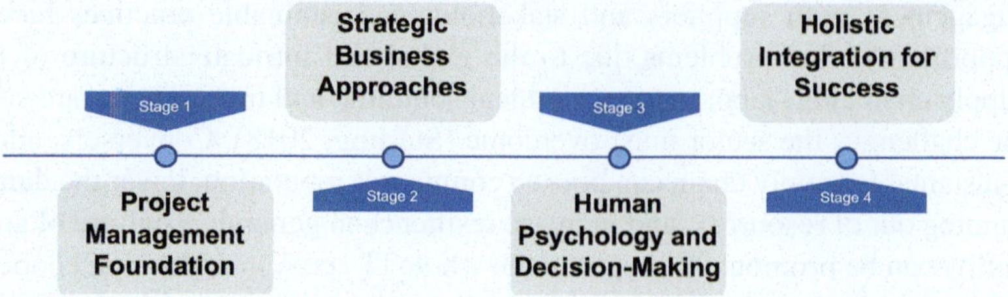

Fig. 10 Integration of project management, business strategies, and human psychology for sustainability

converge is one method that is becoming more and more popular. This essay examines the consequences of this integrated perspective for organizational and environmental well-being, as well as the importance of it in creating sustainable solutions.

1. Project Management, Business Strategies, and Human Psychology at Their Intersection.
   Project Management in Sustainable Initiatives (Johnson & Davis, 2018).
   The foundation for integrating sustainable solutions into organizations is project management. Sustainability goals are guaranteed to be precisely stated, quantifiable, and reachable with the help of an organized strategy (Project Management Institute (PMI), 2017). Iterative development and continual improvement are made possible by project management approaches like Agile and Scrum, which are in line with the dynamic character of sustainability concerns.
2. Business Approaches to Sustainability.
   Beyond merely having good intentions, sustainable solutions necessitate a strategic strategy ingrained in the foundation of the company (Mills, 2011). Key roles are played by business strategies such as circular economy models, sustainable supply chain management, and corporate social responsibility (CSR) efforts. These tactics support long-term economic viability in addition to lessening their negative effects on the environment.
3. Human Psychology in Decision-Making.
   Decisions that determine an organization's position on sustainability are influenced by human psychology. It is essential to comprehend each person's motives, behavioral habits, and cognitive biases inside the company (Johnson & Davis, 2018). Concepts from behavioral psychology, such as incentives and nudges, can be used to motivate stakeholders and staff to perform environmentally friendly actions.
4. The Combination of Holistic Methods.
   Project management, business tactics, and human psychology are all integrated to promote cooperation amongst heterogeneous teams (Sterling, 2018). Effective communication channels are built to ensure that sustainability goals are understood and adopted at all levels of the company (Kolk & Pinkse, 2005). The pos-

sibility of the project being implemented successfully is increased by this synergy.
5. Reaching a Balance Between Economic and Environmental Goals.
   Organizations can achieve a balance between environmental responsibility and economic goals by utilizing holistic techniques. Business strategies that incorporate sustainability considerations can result in cost savings, resource efficiency, and greater competitiveness (Johnson & Davis, 2018). Project management ensures that these initiatives are implemented efficiently and in line with overall corporate goals.
6. Motivating Stakeholders with Behavioral Insights.
   Persuading stakeholders to embrace sustainable practices requires a strong understanding of human psychology (Johnson & Davis, 2018). Understanding the psychological aspects that drive decision-making allows organizations to create solutions that are consistent with individual's values and beliefs. This strategy promotes long-term behavioral change and increases the efficacy of sustainability projects.
7. Adaptability and Agility in the Execution of Projects.
   Because sustainability concerns are dynamic, companies need to be flexible and quick to adjust. Project management concepts can be incorporated into holistic systems to enable iterative changes in response to shifting social, economic, and environmental conditions (Kolk & Pinkse, 2005). This flexibility is critical to the long-term viability of sustainable efforts.

Case Studies

Interface Inc.: One of the best examples of how project management, business tactics, and human psychology can be successfully combined with sustainability is Interface, a multinational modular flooring company. Their 2020 Mission Zero pledge highlights eradicating any adverse environmental impact and is backed by an extensive project management system. Interface has proven how sustainability can be a vital component of corporate success by implementing cutting-edge business techniques like closed-loop manufacturing and biomimicry.

Unilever's Sustainable Living Plan: By combining project management techniques, business tactics, and human psychology considerations, Unilever's Sustainable Living Plan is an example of a holistic approach. Strategic initiatives including sustainable procurement, water efficiency, and waste reduction are all part of the company's commitment to lessening its environmental impact. Through product design and marketing, behavioral insights are used to engage consumers and promote sustainable choices.

Finally, in today's complicated world, establishing long-term solutions requires a comprehensive approach that considers project management, corporate strategies, and human psychology (Hawken et al., 1999). In addition to ensuring efficient project execution, this integrated strategy synchronizes sustainability programs with organizational goals (Porter & Kramer, 2011). The practical advantages of this strategy are illustrated by case studies such as Interface Inc. and Unilever, which highlight the fact that sustainability is not only a corporate responsibility but also a strategic necessity (Johnson & Davis, 2018). As time goes on, companies need to understand how these disciplines are related to one another and work together to solve the environmental problems that modern practices have created.

4.4 Contributing to Shaping Future Practices by Offering a Detailed Understanding of the Historical Context of Energy Footprints in Computing and Their Implications for Environmental Sustainability

The quick development of technology has changed the computing industry and brought about previously unheard-of breakthroughs and discoveries. However, as our reliance on digital gadgets grows, so does the environmental impact of the energy consumed by these technologies (Mills, 2011). Explore the historical background of energy footprints in computing and look at how a thorough knowledge of the subject may help shape future strategies for improved environmental sustainability.

1. The Historical Development of Computing's Energy Footprints.
   Examining the past development of energy footprints is crucial to understanding the present obstacles and opportunities in reducing the environmental effect of computing. In the early days of computers, the emphasis was mostly on functionality and efficiency, with little regard for power usage (Porter & Kramer, 2011). Data centers and electronic devices have a rising carbon footprint due to the expansion of computing capability and corresponding energy requirements. One significant development that brought computing power to both homes and companies was the widespread usage of personal computers in the 1980s (Hawken et al., 1999). Despite being innovative, this democratization of technology increased energy usage. Computing's effects on the environment became more and more obvious as software and applications became more sophisticated and required more processing power.
2. The Development of Data Centers.
   Data centers, which are central locations for the processing, storing, and management of enormous volumes of digital data, first appeared in the late twentieth century. Data centers were significant energy consumers even if they made things more accessible and efficient (Johnson & Davis, 2018). These facilities' growing energy footprint was caused in part by the requirement for constant operation and the requirement for cooling systems to avoid overheating (Hawken et al., 1999). The rise of cloud computing and the digitization of businesses propelled an exponential rise in the quantity and scale of data centers in the twenty-first century. Data centers' environmental impact became a hot topic of

conversation, which sparked initiatives to improve sustainability and energy efficiency in these vital infrastructure elements.

3. Environmental Sustainability Implications.
   Energy footprints in computers have a wide range of environmental consequences. The increase in energy use exacerbates climate change and environmental damage by substantially increasing carbon emissions (Porter & Kramer, 2011). The IT industry must shift to more sustainable practices immediately because the use of nonrenewable energy sources exacerbates these issues. Considering the full life cycle of electronic equipment is necessary to address the carbon footprint of computers, in addition to focusing on energy consumption reduction. The environmental impact is influenced by the extraction of raw materials, production methods, and the disposal of electronic waste. For true sustainability to be achieved, therefore, a comprehensive strategy that goes beyond energy efficiency is essential.
4. Contributions to Forming Upcoming Procedures.
   Future practices in computing are greatly influenced by a thorough understanding of the historical background of energy footprints (Porter & Kramer, 2011). By studying how computing technology has developed and the effects it has had on the environment, we can make informed strategic decisions and direct the creation of long-term solutions (Hawken et al., 1999). In the domain of knowledge-based decision-making, a thorough understanding of past energy consumption trends provides decision-makers with the information they need to develop knowledge-based strategies (Kolk & Pinkse, 2005). Organizations are better equipped to prioritize energy-efficient technology when adopting new ones because they have a better grasp of the variables that have contributed to the increase in energy footprints (Porter & Kramer, 2011). This educated decision-making approach serves as a foundation for guiding enterprises toward a more sustainable path, aligning technical investments with environmental concerns. The optimization of computing infrastructure has its roots in historical background, which provides a solid foundation for improving existing systems (Johnson & Davis, 2018). By carefully examining historical patterns in energy usage, companies can identify areas that are in dire need of repair (Porter & Kramer, 2011). Equipped with this understanding, specific actions can be put in place to improve the effectiveness of data centers and computing infrastructure (Hawken et al., 1999). In addition to lowering energy footprints, this optimization puts businesses in a strategic position to handle the changing needs of sustainable computing (Kolk & Pinkse, 2005). Informed decision-making and infrastructure optimization work together to create a powerful synergy, moving companies toward a future in which technology and environmental stewardship are seamlessly integrated.

In summary, how future strategies for improved environmental sustainability are shaped depends critically on a thorough grasp of the historical background of energy footprints in computing. We can steer toward a future in which the IT industry not only satisfies its functional needs but also functions in harmony with the environment by recognizing the way technology is evolving and the effects it has on the environment. Decision-makers, scholars, and the public can work together to create a more sustainable digital future by following the path laid down by the lessons learned from the past.

5 Conclusion

Examining and comprehending energy footprints in computing from a multidisciplinary perspective offers priceless insights into the intricate mechanisms influencing patterns of energy consumption. The IT industry's organizational and individual decision-making processes are clarified by combining viewpoints from project management, business analysis, and behavioral psychology. These perspectives also highlight key factors influencing energy efficiency. Historical analyses provide based on fact understanding of how energy footprints have changed over time, providing a basis for well-informed responses to environmental issues. To solve environmental concerns in the IT sector and promote sustainable practices, strategic solutions informed by interdisciplinary viewpoints must be applied. Collaboration, adaptability, and teamwork are key components of this approach. Through the utilization of interdisciplinary expertise and historical context, actionable measures can be implemented to mitigate ecological footprints and promote the uptake of ecology.

References

Ajzen, I. (1991). The theory of planned behavior. *Organizational Behavior and Human Decision Processes, 50*(2), 179–211.

BABOK Guide. (2015). *A guide to the business analysis body of knowledge (BABOK GUIDE)*. International Institute of Business Analysis (IIBA).

Beloglazov, A., & Buyya, R. (2010). *Energy-efficient resource management in virtualized cloud data centers*. In 2010 10th IEEE/ACM international conference on cluster, cloud and grid computing (CCGrid) (pp. 826–831). IEEE.

Brown, L., & Vergragt, P. (2018). *Time and the technosphere: The law of time in human affairs*. Routledge.

Brown, C., & White, D. (2019). Environmental impact reduction: Integrating LCA into business strategies. *Journal of Sustainable Business, 15*(2), 123–140.

Chen, Y., Li, S., Li, T., & Zhu, Y. (2020). Exploring the impact of green information technology maturity on organizational energy efficiency: An empirical study. *Information Systems Frontiers, 22*, 187–203.

Curran, M. A. (2016). *Life cycle assessment handbook: A guide for environmentally sustainable products*. Wiley.

Davis, K. (2017). Human behavior and energy efficiency: Challenges and opportunities. *Energy Research & Social Science, 28*, 58–69.

Ellen MacArthur Foundation. (2021). *The circular economy*. https://www.ellenmacarthurfoundation.org/circular-economy/overview/concept

Gifford, R. (2011). The dragons of inaction: Psychological barriers that limit climate change mitigation and adaptation. *American Psychologist, 66*(4), 290–302.

Green, M., & Brown, N. (2018). The role of project management in sustainable business practices. *Journal of Sustainable Project Management, 5*(2), 78–92.

GWEC. (2020). *Global wind report 2020*. Global Wind Energy Council. https://gwec.net/reports/global-wind-report-2020/

Hawken, P. (2017). *Drawdown: The most comprehensive plan ever proposed to reverse global warming*. Penguin.

Hawken, P., Lovins, A., & Lovins, L. H. (1999). *Natural capitalism: Creating the next industrial revolution*. Back Bay Books.

Johnson, R. L., & Davis, C. M. (2018). Behavioral psychology in environmental conservation. *Journal of Environmental Psychology, 25*(3), 345–356.

Johnson, C., & Patel, R. (2019). *Project management for sustainable development in the IT industry*. Springer.

Johnson, E., & Turner, F. (2021). Sustainable supply chain management: A business analysis approach. *Sustainability, 13*(5), 2654.

Kahneman, D. (2011). *Thinking, fast and slow*. Farrar, Straus and Giroux.

Kahneman, D., & Tversky, A. (1979). Prospect theory: An analysis of decision under risk. *Econometrica, 47*(2), 263–291.

Kerzner, H. (2017). *Project management: A systems approach to planning, scheduling, and controlling*. Wiley.

Kolk, A., & Pinkse, J. (2005). Business responses to climate change: Identifying emergent strategies. *California Management Review, 47*(3), 6–20.

Koomey, J. G. (2007a). Worldwide electricity used in data centers. *Environmental Research Letters, 2*(1), 014002.

Koomey, J. G. (2007b). *Estimating total power consumption by servers in the US and the world*. Stanford University.

Kueng, P., & Lyytinen, K. (2016). Beyond buzzwords: A taxonomy of blockchain systems. *MIS Quarterly Executive, 15*(2).

Lauesen, S. (2019). *Software requirements: Styles and techniques*. CRC Press.

McDonagh, P., Prothero, A., & Ahmed, S. (2019). Sustainability marketing research: Past, present, and future. *Journal of Marketing Management, 35*(1–2), 105–142.

Mills, E. (2011). The carbon footprint of information and communication technology. *Journal of Industrial Ecology, 15*(5), 805–820.

Porter, M. E., & Kramer, M. R. (2011). Creating shared value. *Harvard Business Review, 89*(1/2), 62–77.

Project Management Institute (PMI). (2017). *A guide to the project management body of knowledge (PMBOK guide)*. Project Management Institute.

Ross, J. W., Weill, P., & Robertson, D. C. (2006). *Enterprise architecture as strategy: Creating a foundation for business execution*. Harvard Business Press.

Schaltegger, S., Lüdeke-Freund, F., & Hansen, E. G. (2012). Business cases for sustainability: The role of business model innovation for corporate sustainability. *International Journal of Innovation and Sustainable Development, 6*(2), 95–119.

Schwalbe, K. (2018). *Information technology project management*. Cengage Learning.

Shtub, A., Rosenwein, M., & Globerson, S. (2018). *Project management: Processes, methodologies, and economics*. Pearson.

Smith, J. A., & Johnson, R. B. (2018). *Business analysis techniques: 99 essential tools for success*. BCS, The Chartered Institute.

Sterling, T. (2018). *High-performance computing: A history*. MIT Press.

Sun, Y., Du, J., Chai, Y., Feng, X., & Jin, H. (2018). Understanding the impact of blockchain application on corporate social responsibility: A resource-based view. *Sustainability, 10*(10), 3548.

Sustainable Apparel Coalition. (2021). *About us*. https://apparelcoalition.org/about/

Sustainable Business Council. (2022). *Best practices for sustainable supply chain management in the IT industry*.

Thakkar, J., & Laverty, J. (2019). Sustainable fashion in the mainstream: An exploration of Patagonia's positioning as a sustainable fashion brand. In M. Ali & K. Vequist (Eds.), *Handbook of research on business ethics and corporate responsibility* (pp. 298–314). IGI Global.

Thaler, R. H., & Sunstein, C. R. (2008). *Nudge: Improving decisions about health, wealth, and happiness*. Yale University Press.
Tversky, A., & Kahneman, D. (1974). Judgment under uncertainty: Heuristics and biases. *Science, 185*(4157), 1124–1131.
Walker, D. H. T. (2015). *Sustainable project management*. Kogan Page.
Wijen, F., & Ansari, S. (2007). Best of both worlds: Using existing capabilities and creating new competencies in facing changed environmental conditions. *Organization Studies, 28*(4), 489–511.

Providing Decent, Affordable, and Sustainable Housing: Analysing Environmental Impacts of Family Houses Built with Conventional and Unconventional Building Materials

Aron Szennay, Zoltán Major, Róbert Horváth, Herta Czédli, and Cecília Szigeti

1 Introduction

The construction industry and buildings themselves play a significant role in climate change due to their energy consumption, construction materials, and design. Green buildings have emerged as a crucial solution to mitigate the environmental impact of buildings on climate change (Jia & Wang, 2012). Green buildings are designed to be energy-efficient, using sustainable materials and technologies to reduce energy consumption and minimise greenhouse gas emissions, and they not only provide healthier indoor environments for occupants but can also contribute to reducing air pollutants and improving overall public health (Allen et al., 2015). The construction sector's resource consumption and environmental impacts highlight the necessity of transitioning towards more sustainable building practices (Wu et al., 2019), while

A. Szennay (✉)
Budapest Business University, Budapest LAB Office for Entrepreneurship Development, Budapest, Hungary
e-mail: szennay.aron@uni-bge.hu

Z. Major · R. Horváth
University of Győr, Győr, Hungary
e-mail: majorz@sze.hu; horvath.robert@sze.hu

H. Czédli
University of Debrecen, Debrecen, Hungary
e-mail: herta.czedli@eng.unideb.hu

C. Szigeti
Budapest Metropolitan University, Sustainability Studies Institute Budapest, Budapest, Hungary
e-mail: cszigeti@metropolitan.hu

S. S. Muthu (ed.), *Energy Footprint and Sustainability*, Environmental Footprints and Eco-design of Products and Processes, https://doi.org/10.1007/978-3-031-63057-6_6

the adoption of sustainable practices can significantly impact society's progress towards Sustainable Development Goals (Pitt et al., 2009).

The construction industry plays a crucial role in contributing to and achieving the Sustainable Development Goals of the Agenda 2030 adopted by the majority of member states of the United Nations (2015) through various initiatives and practices. Climate change mitigation measures in the construction sector can have significant social and economic value. They enable the achievement of four Sustainable Development Goals and reinforce seven others. Furthermore, they are directly linked to the achievement of five SDGs, namely SDG 3 (good health and well-being), SDG 7 (affordable and clean energy), SDG 8 (decent work and economic growth), SDG 11 (sustainable cities and communities), and SDG 13 (climate action) (Mirasgedis et al., 2024). Alternative, innovative solutions from the construction industry can make a significant contribution to achieving the SDGs (Omer & Noguchi, 2020). Hauashdh et al. (2024) demonstrated the significant role of sustainable and efficient building maintenance operations in existing buildings in achieving several of the UN Sustainable Development Goals. Maintenance decisions for ageing buildings based on environmental impact assessments are consistent with SDG 13 (climate action). It also supports the achievement of SDG 11 (sustainable cities and communities) by emphasising the importance of well-maintained buildings. In addition, using emerging technologies in building maintenance operations contributes to SDG 9 (industry, innovation, and infrastructure) by promoting the uptake of innovative technologies. Research on smart city solutions shows that responsible development/implementation of smart city solutions and technologies can contribute to progress towards the SDGs. The literature focuses mainly on SDG 11 (sustainable cities and communities), SDG 12 (responsible consumption and production), SDG 7 (affordable and clean energy), and SDG 6 (clean water and sanitation) (Sharifi et al., 2024).

The chapter aims to gain a better understanding of the environmental impacts of housing in Hungary, analysing some dwellings typical in Hungary, namely (1) a new two-storey family house with a gable roof, (2) a new family house made of timber, and (3) an energy upgrade of a typical family dwelling built between 1950 and 1990, called "Hungarian cube". This sample can be considered appropriate as these options represent the options of living in a single-family dwelling in Hungary and even in Central and Eastern Europe.

The chapter is structured as follows: First, we provide a brief overview of the housing stock and its trends in Hungary. Then, the environmental impacts of the construction industry and the real estate sector are analysed, with special respect to its sustainability aspects. At the end of the literature review, possible ways of environmental impact measuring are presented, highlighting some ways how to make housing more sustainable. It is followed by a brief description of the methodology and the results of the analysis. At the end of the chapter, the results are discussed, recapitulating their policy and practical implications, and a summary is also provided.

2 Theoretical Background

2.1 Housing Stock in Hungary

The results of the Hungarian Census 2022 show that while the population has been steadily declining since the 1980s, the housing stock is constantly increasing (HCSO, n.d.). It means that the population decreased from 10.7 million in 1980 and 9.9 million in 2011 to 9.6 million in 2022, while the number of dwellings has increased by more than 1 million from 3.54 to 4.58 million in that period. People in Hungary tend to prefer living in houses as three out of four (72.1%) people live in houses, while the EU average is slightly higher than one in two (52.7%) (Eurostat, n.d.). Furthermore, ownership of dwellings is rather typical, like in other former socialist EU member states. Eurostat (n.d.) shows that the dwelling ownership rate of Hungary is the third highest (91.7%) in the EU, after Romania (95.3%) and Slovakia (92.9%), while the EU average lies at 69.9%.

According to the latest Hungarian Census of 2022, 87.5% of the dwellings are occupied. The typical building material of the outer wall is brick, stone or other manual walling element (64.9%). In contrast, industrialised building methods (panel, middle or large block, cast concrete) used overwhelmingly in the socialist era have a share of 18.0% in the housing stock. Alternative or even traditional building materials like adobe or mud, however, are not used contemporarily. Although they were used as the material of the outer wall in the case of 13.4% of dwellings, 87.9% of them were built before 2001 (Table 1).

There are three dwelling types with a major share in the Hungarian housing stock: (1) the panel buildings, (2) middle or large blocks, or even cast concrete buildings, and (3) the so-called Hungarian cube. All of them were built overwhelmingly in the socialist era between 1947 and 1989, but while the first two are condominiums built with industrialised technology, the latter is a single-family house. Hungarian cubes are single-storey houses with a pyramid-shaped roof and a floor area of 8 × 8 or 10 × 10 m, and are built of bricks. As approximately 850 thousand of them were built between the 1960s to the early 1990s, this architecture is the

Table 1 Number and share of dwellings by the occupancy and the material of the outer wall

	Occupied conventional dwellings		Unoccupied conventional dwellings	
	n	%	*n*	%
Brick, stone, manual walling element	2,589,130	64.6	383,178	67.0
Panel	514,655	12.8	54,921	9.6
Middle or large block, cast concrete	226,596	5.7	29,179	5.1
Adobe, mud, etc., with a solid basement	522,060	13.0	92,601	16.2
Other	156,100	3.9	12,118	2.1
Total	4,008,541	100.0	571,997	100.0

Source: HCSO (n.d.)

most characteristic house of the Hungarian countryside (Szécsi, 2021). Although they were considered modern comparing the traditional village houses providing a more civilised standard of living (see, for example, Tamáska, 2021), their energy attributes are rather poor due to the lack of additional insulation and inefficient heating (Matolcsy, 2020). As the annual energy demand of a Hungarian cube is around 400–550 kWh/m^2, their refurbishment can provide significant savings both in terms of environmental impact and financial costs (Szécsi, 2021). An energy renovation can provide significant savings in both monetary and environmental terms. According to the results of the Hungarian National Energy Efficiency Institute, refurbishment to the level of near-zero energy demand could mean energy savings of ~31.2 PJ, 4.5-fold of the annual target of the National Energy and Climate Plan (Illésné Szécsi et al., 2021; Szécsi, 2021). Using life cycle analysis, Szalay és Gröller (2021) showed that an optimised energy upgrade (i.e., insulation, replacing windows and doors, using a condensation furnace, using LED lights) has lower lifetime costs in the 30-year time frame than the baseline scenario.

2.2 *Environmental Impact of the Construction Industry and the Real Estate Sector*

Life cycle research in the construction industry is a major focus for green and sustainable processes and techniques, with a particular focus on energy consumption and environmental impacts (Sandanayake, 2022). Buildings and construction are a major source of resource consumption and environmental emissions, and studies have demonstrated the need to identify emission reduction practices (Buyle et al., 2013; Sandanayake, 2022; Zhang et al., 2013). According to various statistical analyses and preliminary studies, buildings account for one-sixth of the world's freshwater withdrawals, one-quarter of the timber harvested, and two-fifths of the materials used, indicating intensive resource consumption (Chau et al., 2012; Sandanayake, 2022). Improving the environmental performance of buildings is an important issue in today's densely populated cities. Measured data show that air pollution from buildings is one of the largest contributors to many sources of pollution (Zhang et al., 2013). Buildings are responsible for over a third of global energy-related carbon dioxide (CO_2) emissions in the twenty-first century (Ürge-Vorsatz & Novikova, 2008). The main conclusion is that significant, low-cost reductions in energy-related CO_2 emissions in the construction sector can be achieved in the coming years (Ürge-Vorsatz & Novikova, 2008). Literature suggests that if this potential were realised, CO_2 emissions from buildings would remain constant between 2004 and 2030. These stabilisation levels (if achieved by all other sectors) could offset a temperature increase of about 3 °C (Ürge-Vorsatz & Novikova, 2008).

Creating the foundations for a viable and sustainable future requires a harmonious balance between the natural environment and humanity in all sectors (S. Cao et al., 2023; Lupton & Samy, 2022).

In the modern world of the twenty-first century, population explosion and industrial revolution have posed unprecedented challenges to human society in the relationship between the economy and the environment, and environmental protection has become a new focus area of modern civilisation in the course of socio-economic development (Berrone et al., 2013; Kocsis, 2014; Cao et al., 2023).

Building the present means shaping the future. Important transformations are needed in the building economy, as available data show that building-related emissions account for 37% of the total global carbon footprint (United Nations Environment Programme, 2023). We aim to raise awareness of the carbon footprint of the construction industry (construction, maintenance, and demolition of buildings, transport processes), as it is the largest global consumer of resources (Huang et al., 2018; Solís-Guzmán et al., 2014; World Economic Forum, 2016).

Effective action is needed to achieve and meet domestic and international sustainability targets and to ensure a low-carbon transformation of the building sector. The future cannot be viable without effective cooperation and joined-up thinking by many actors in the sector. The construction industry plays a significant role in improving the overall quality of life of society in our country and the world. Through its activities, the sector contributes to ensuring living standards, economic stability, comfort, and design value to achieve and meet national and international sustainability goals and to ensure the low-carbon transformation of the construction industry (Alawneh et al., 2018; AlSanad, 2015; Stanitsas et al., 2021). The concept of "green building" has emerged as a response to negative and environmentally damaging impacts, as a strategy to reduce carbon emissions and promote more environmentally friendly building practices (Jaradat et al., 2024).

Effective collaboration between experts from different disciplines can deliver a carbon-neutral vision for the built environment. The common goal, nationally and internationally, is a *transition to an energy-efficient, regenerative, and equitable built environment* (World Green Building Week, 2023).

Looking at the construction aspects of international and Hungarian policy frameworks (Beleznay, 2021; Zalatnay, 2016), we can conclude that (1) buildings are among the largest domestic emitters of CO_2 and energy consumers; (2) transport related to construction projects puts additional pressure on the environment; and (3) the decarbonisation targets of the construction industry and the efforts to achieve them contribute to the national commitment. The building industry is a vital element of any economy but has a significant impact on the environment (Akadiri et al., 2012).

The weight of the construction sector in the national economy in terms of gross value added is around 4% and has been increasing since 2016, reaching 6.5% in 2019 (Zalatnay, 2016). The number of construction companies is 4.9%, and the number of employees is 7.6%. Although average monthly gross earnings in construction have increased in recent years, construction is at the bottom of the ranking of sectors, 22% below the average. "The growth in construction output has had a positive impact on the propensity of construction companies to invest in recent years, which has been steadily increasing since 2016" (Beleznay, 2021).

"The increase in high-rise construction has been driven by residential buildings (through supportive subsidies), while non-residential construction has been boosted by the construction and renovation of industrial, commercial, educational (mainly university) buildings and sports halls" (Beleznay, 2021). The built environment and human interventions are taking up more and more land, resulting in shrinking ecosystems (He et al., 2023). Only a small area of Hungary contains natural ecosystems or areas suitable to produce ecosystem services (forests, tree plantations, grasslands, vineyards, orchards) (Zalatnay, 2016). The problem is that these areas are fragmented, and fragmentation is increasing. Ecosystem services are being compromised by the fragmentation of previously contiguous, coherent habitats and the elimination of ecological corridors. Service providing units (SPUs) can vary from service to service and need to be considered in the planning practice. For urban ecosystem services, it is important to highlight that site specificity may be more important than for some other services: for example, the local climatic context of a city will determine the importance of the microclimate regulating the potential of trees for its population, and therefore the scale of the built-up/overground cover should be interpreted in this context (Beleznay, 2021). Artificial interventions that directly or indirectly alter natural systems are built-up areas, fields, pastures, oil and gas production, mining and quarrying, electricity generation, roads, railways, power lines, electricity infrastructure, logging, human interventions, reservoirs, and air pollution (Beleznay, 2021). Although most of the human interventions take place outside of our continent, both Europe and Hungary have seen an increase in built-up and paved areas over the last decade. Hungary is one of the EU Member States with the largest increase in artificial land cover, with a 14% increase between 2009 and 2015 (Beleznay, 2021). Artificial land cover is increasing in Hungary, while the amount of green space per capita owned by municipalities is steadily decreasing (Beleznay, 2021). There is almost no landscape in our country that is free from the effects of socioeconomic activity. This poses a threat to ecosystems, natural habitats, biodiversity, and the water cycle. **The built environment that surrounds us is pushing us towards greater resource use in all its elements.** *Hungary's performance over the last decade has been extremely poor in three main areas (land use, material flows, and anthropogenic climate change), and one of the most destructive in the EU in terms of natural capital use. Land use in* the *country is most affected by urbanisation processes, the predominance of infrastructure, real estate and construction development, and the proportion of greenfield and brownfield investment* (European Commission. Directorate General for Economic and Financial Affairs, 2023). The greatest health risk from climate change are heat waves, quantified in terms of rising mortality rates. Build-up can further increase the urban heat island effect, which can create health hazards in built-up urban environments, with or without the onset of heat waves. Artificial land cover and further disproportionate increase and expansion of impervious surfaces will increase the risk of flash floods, especially if climate change also increases the number of high precipitation periods. In our cities, the development of blue-green infrastructure could help alleviate this problem (Foster et al., 2011), but it is about "breaking up" concrete, not creating additional paved surfaces. The built environment, the man-made space that integrates living, working, and recreation, includes physical structures, supporting infrastructure, and the planned and managed urban green and blue spaces where people carry out these activities (He et al., 2023).

Artificial land is being created partly because of suburbanisation and urban sprawl due to weak legal frameworks, and partly because of infrastructure development, international and national transport, and infrastructure networks, which are playing a major role in EU and national strategies (Vasárus & Lennert, 2022). Both processes are driven by economic growth. While at the macro-regional level the focus is on trans-European transport and energy networks—TEN-T and TEN-E—the motivation is at the national level on access to jobs and settlements and on transport routes to and from construction product supply chains. The planned transformation of the energy sector will result in further developments of the electricity grid and the renewable energy infrastructure in the coming decades. Both infrastructure networks and logging related to construction and renewable energy (biomass) will need to ensure the maintenance and enhancement of forests as natural sinks of CO_2, the restoration of degraded forest areas and other ecosystems, the interconnection of a mosaic system that, in addition to economic benefits, can ensure the sustainability of biodiversity, and many other positive environmental impacts (soil and water protection) and social impacts (health, air quality) (European Energy Agency, 2023). Material flows involve the extraction, transformation, and waste of natural resources. The depletion of available natural resources also raises the problem of scarcity and substitutability in the twenty-first century. Resource use requires professional competence, continuous communication, coordination, and trade-offs while maintaining the priority of environmental and economic sustainability. Those municipalities that plan consciously and outline scenarios that improve the quality of life of their inhabitants will be favoured. The economic growth of the last decade has been accompanied by a massive expansion in the use of materials. From 92.1 million tonnes extracted in 2010, the amount of material used has increased by one and a half times to 138.6 million tonnes in 2018. Hungary was the most efficient country in the region in terms of both domestic material consumption (DMC) per capita and resource productivity until the mid-2010s but has been regressing in recent years (OECD, 2023). The built environment is a major resource-intensive sector. The construction industry uses around 50% of all extracted materials and is responsible for 35% of waste generation, making it a key sector in the circular economy (Beleznay, 2021). Even though many actors in the construction industry are implementing solutions to reduce their carbon footprint, the sector continues to struggle with the application of full, deep interventions, and a true circular mindset in business models. Given the clear correlation between the growth of material use and the decline of our natural resources, it is a minimum requirement that the growth of material use in the construction sector should be slower than the growth of the associated national product (GDP) (relative decoupling) (OECD, 2023). It is also necessary to examine whether it is possible to decouple economic development and material consumption in absolute terms (absolute decoupling, material use declines or stagnates while GDP grows or at least changes more than material use). While in both cases the efficiency of resource use increases, only the latter results in a reduction of pressure on the environment. In the previous period,

2005–2012, Hungary achieved absolute decoupling for the economy, so that the latter can be set as the goal for resource protection and sustainable use, and all policy, legal, and economic incentives should be directed towards achieving it. One of the foundations for creating a circular economy is to make the best use of the commodity chains within the country. At present, Hungary's imports of goods and services, which account for 81.2% of GDP and exports for 83.3%, are above the EU average, making it one of the less efficient circular economies. Countries that have adopted a circular construction strategy and fostered an efficient value chain network in the construction economy—from mining, through materials and structures manufacturing, design and professional services, developers, contractors and their subcontractors, suppliers, and traders, to property managers and specialised service providers—both through regulation and incentives and by creating economies of scale, are among the leaders in these indicators. The sustainable supply of primary and secondary raw materials, especially critical raw materials for key sectors, is the basis for the country's economic development (Talapin & Braun, 2020). According to the European Commission's 2020 assessment of global supply and EU trade, Hungary is affected by the raw materials bauxite, bentonite, coking coal, fluorite, gallium, kaolin clay, manganese, perlite, and quartz sand (Beleznay, 2021). Aggregates, which account for the largest amount of material use, are critical for environmental sustainability due to the negative environmental impacts of extraction and secondary aggregate sites, the potential depletion of sand and gravel reserves and irreversible damage through over-mining (Kisku et al., 2017).

The development of a secondary aggregate market is key and could be competitive, but environmental, economic, and governance disincentives need to be addressed. The first is the difficulty in product quality and environmental testing of secondary additives, the second is the lack of confidence in secondary products, and the last is the lack of coherence and coordination between different authorities with different interpretations and interests. With the development of the construction sector, the volume of construction and demolition waste, which represents a significant part of the waste volume, is also increasing. In Hungary, because of the lack of attention paid to demolition, selective collection by waste groups is not implemented, so the quality of recycling of construction and demolition waste is limited, and it is mainly used in road foundations and backfill. The waste sector contributed 5% to Hungary's total GHG emissions in 2017, but the associated transport significantly increased this. With a 35% rate of preparation for reuse and recycling, Hungary is below the EU average of 47%. Hungary's opportunity lies in building a quality supply system rather than quantity production, and in developing the supply side of low-carbon construction through innovation and research and development. Furthermore, it is important to create short supply chains that minimise surpluses within the system and reduce emissions by reducing transport routes within the country. In addition, human and natural resources, especially those that will be in short supply in the future, need to be prioritised and increased (or at least maintained at a level that meets the needs of the system). The medium-term strategy for the improvement of efficiency and performance of the construction sector and the sustainable development of the built environment in Hungary reflects the problems and needs of the sector and formulates proposals for the qualitative improvement and sustainability of the built environment, as well as for the efficiency, productivity, and stability of the sector (Beleznay, 2021). The most important competencies to be developed both in Hungary and in the EU are (1) green and energy-efficient construction (knowledge of materials, technologies, and processes); (2) digital skills (digital technologies in construction and investment management); (3) new and innovative design technologies; and (4) use of new and innovative materials. The third major factor is the prevention and reduction of climate change because of anthropogenic processes. Several national strategies related to climate change were presented in early 2020 (National Energy Strategy, National Clean Development Strategy, Hungary's National Energy and Climate Plan, and First Action Plan on Climate Change), which focus on the decarbonisation of the energy and transport sectors to achieve the 2050 goal of a carbon-neutral Europe and Hungary.

Hungary is among the EU's low emitters of greenhouse gases per capita (European Commission. Directorate General for Economic and Financial Affairs., 2023). However, when looking at the components by Kaya identity, while emissions are decreasing overall, the combined effect of GDP per capita and population change is larger than the effect of decreasing energy intensity and carbon intensity. In addition to decarbonising energy production, improving energy efficiency can make an effective contribution to the targets. Buildings should play a key role in this, given that buildings are among the largest domestic energy consumers and CO_2 emitters. Residential buildings account for 35% of total final energy consumption and public buildings for 28%. The Long-Term Renewal Strategy (HTFS) aims to transform the national stock of residential and non-residential buildings, both private and public, into a near-zero energy and decarbonised building stock by 2050 by promoting and accelerating renovation (European Commission, n.d.). At present, the building stock is rather obsolete, 95.5% of the buildings are woefully below the mandatory KNE energy certification level from 1 January 2021, and the legal and economic incentive framework does not promote energy efficiency. Reducing the energy consumption of buildings is important not only because it currently accounts for nearly 40% of our energy-related CO_2 emissions, but also because it is one of the few sectors that can achieve net zero or very close to net zero results independently of other sectors (commission.europa.eu). In a tight time frame, full electrification and the provision of zero-carbon electricity from renewable and nuclear sources cannot solve everything, as heating accounts for a significant part of energy consumption in buildings (Cao et al., 2016). The HTFS, TNM Regulation 7/2006 (24 May 2006) on the definition of the energy performance of buildings (TNM Regulation) and the programs and funding instruments supporting building renovation can support climate protection if much stronger criteria are set than currently exist. In addition to buildings, the energy-intensive, hard-to-abate industries of the construction sector, steel and cement production, are also potential mitigation options. For steel and concrete, which are the materials largest by quantity used in the construction and maintenance of buildings, the emissions from their production exceed the carbon load from all other stages of the life cycle. Furthermore, GHG emissions from materials extraction, construction products, building construction, and renovation, which account for 5–12% of national GHG emissions, can be reduced by 80% by increasing material use efficiency (Interreg Europe, n.d.).

Straw bale architecture can be built with the lowest energy use of all building energy aspects.

- In terms of raw material production, straw bales use 30 times less energy as polystyrene, for example.
- The great advantage of straw bales is that they are residues from harvesting cereal grains, presently the largest food group for human consumption, as an agricultural by-product, and they can be found wherever there are cereal fields. Thus, transport to the construction site requires minimal energy.
- Heating and cooling buildings is the largest factor of energy use in the life cycle of a building. Straw bales can be used to create a certified passive house. Good thermal insulation is a prerequisite for low ecological impact.
- Because of the low heat demand of straw bale buildings, it is possible to use alternative renewable energy sources that would previously have been impossible or would have been cost-prohibitive in poorly insulated homes. The use of rocket stoves and solar energy in these buildings also offers the possibility of providing heating affordably.
- The removal and disposal of waste from the demolition of buildings requires additional energy costs and creates an environmental burden due to the landfilling of materials. The straw itself is biodegradable. It only needs to be exposed to moisture, and the cellulose-degrading bacteria do their job. The storage, disposal, or subsequent treatment of straw, therefore, have no environmental impact.

2.3 *Measuring the Environmental Impact of Housing*

The impact of technological issues in construction on the ecological footprint has attracted considerable interest in the literature. Alba-Rodríguez et al. (2022) analysed eight practical projects in Spain. Material use can be reduced by 80%, depending on the recycling and recovery of materials. The biggest impact is in the foundation and structure chapters, responsible for 30% of the total ecological footprint. With the energy improvements implemented, CO_2 emissions can be reduced by 60% on average. Footprint calculations (LCCFs) can provide a useful assessment metric for analysing and comparing the environmental impact of buildings (Schwartz et al., 2018). Marrero et al. (2024) outline three options:

- Replacing a mesh slab with concrete blocks for a slab with a retrievable casing slightly increases the risk of meeting the deadline due to the increased duration of work from height, but this is offset by a reduction of more than 30% in most other indicators.
- Replacing the brick partitions with drywall increases the project cost and WF due to the installation of steel framing, but these increases are offset by a significant reduction in CDW of 70% and 40% for the other indicators.

- The largest reduction is due to the change in the design of the building facade, from clad brick to covered facade brick, eliminating the single-layer coating, resulting in a nearly 38% reduction in the environmental impact of the work unit.

Net zero requirement for the whole life cycle can be achieved for both prefabricated brick and timber buildings by installing photovoltaic panels with the necessary dimensioning (Szalay et al., 2022).

Analysis of the environmental impact of buildings allows us to identify and even replace the materials with higher impacts. For example, the findings of Schwartz et al. (2018) show that three materials had a higher ecological footprint in all the analysed cases: (1) cement, (2) steel, and (3) ceramic materials. Szennay et al. (2021) have also found that the majority of the ecological footprint of SFHs is determined by some materials, including concrete, reinforcing steel, bricks, and roof tiles.

Since the environmental impact of construction is largely determined by the volume of cement used, scholars aim to optimise its usage or even find substitutes as binders in the case of producing concrete. Szép et al. (2023), for example, have found that a 10% reduction of the ecological footprint can be achieved by using concrete recipes with cement substitutes, taking into consideration resistance aspects as well. For example, cellular concrete powder (CCP), a by-product of cellular concrete block production, can be used as a substitute for metakaolin, but their combined usage also does not affect the performance of concrete significantly (Gyurkó et al., 2019). Analysing the impact of using recycled aggregates in concrete, Mengistu and Nemes (2024) have found that even a 67% replacement level holds regardless of the concrete strength class, which can also significantly reduce the environmental impact of concrete. Furthermore, recycled aggregates, following proper proportioning and mixing, can be used even for producing high-strength self-consolidating concrete (HSSCC) as well (Abed et al., 2022; Abed & Nemes, 2019). In the case of building residential buildings with lightweight structures (e.g. timber, straw, plasterboard, etc.), the volume of concrete built-in can be significantly reduced or even omitted. Using a concrete point base can be an effective substitute for conventional concrete foundations, while evidence shows that ground bolts can be used as foundations in the case of particular soil types.

Results of Szalay et al. (2022, p. 13001) show that "the ratio of embodied impacts and the operational impacts without solar energy are about 1:2 for wooden buildings and 1:1.5 for brick buildings". A systematic review by Schwartz et al. (2018), however, concluded that the average embodied, operational-related and demolition-related CO_2 of 251 case study buildings is responsible for 24%, 75%, and 1%, respectively, of life cycle carbon footprint (LCCF) calculations. Yang et al. (2020), however, found that 81.4% of the total ecological footprint of a residential building (Xining, China) occurred in the pre-usage stage, which is a significantly higher figure than those of the abovementioned ones.

The total impact, however, can be reduced by 80% depending on the recycling and recovery potential of the materials. The foundation and structure chapters have the largest impact, with 30% of the total EF. Energy improvements implemented reduce CO_2 emissions by an average of 60% (Schwartz et al., 2018).

Based on the theory of ecological footprint, a study by Yang et al. (2020) established a building life cycle assessment model that takes residential buildings in Hsingin as the object of research; the ecological footprint of residential buildings in Hsingin was calculated using the advanced ecological footprint model, the "domain hectare method". The aim is to understand the impact of buildings on the environment and address pollution at source. The results show that the ecological footprint of a building is 1.52 times its ecological carrying capacity. The ecological deficit should be reduced by improving the sustainable use of materials to promote sustainable development in the building industry (Yang et al., 2020).

As mentioned above, innovative solutions can reduce the environmental impact of building materials (e.g. concrete), and some techniques are also available, which could result in buildings with lower embedded impacts. Using non-conventional building materials is, however, also an adequate answer if their performance and other attributes are appropriate. A case study of a straw bale house located in Bavaria (Germany) showed that the walls provide excellent living conditions as they moderate the internal humidity and smooth out extreme temperatures (Ashour et al., 2011). González (2014) compares the environmental impacts of walls made of (1) fired bricks, (2) concrete blocks, (3) straw-clay blocks, (4) straw bales, and (5) half straw bales. The results show that non-conventional methods have both a significantly lower environmental impact and superior thermal conductivity compared to conventional materials. These results are confirmed by Li et al. (2021), as wood-structure and light-steel structure rural houses with straw bale walls have 96.75 and 76.92% lower net carbon emissions compared with the reference. Although these values consider the carbon offset of natural building materials, rural houses built with non-conventional methods have a better environmental performance even without accounting for the carbon sink effect. The importance of carbon offsetting is highlighted by the results of Sodagar et al. (2011) since it provides a net positive environmental effect after considering carbon sink, which means that it is environmentally beneficial to build a straw bale house.

3 Methodology

The chapter follows the standardised ecological footprint calculation approach proposed by Szigeti et al. (2023). Thus, the calculation aims to determine the embodied carbon of building materials and buildings first, in which the amount of carbon is converted to ecological footprint. The calculation uses four steps based on an unpriced work breakdown structure (WBS) which contains all elements of the projects with their quantity in a natural measure, for example, in pieces in the case of a particular brick, or cubic metres in the case of a specific concrete. In some cases, however, an element represents a work process outsourced to a contracted partner, but it is typical only for the prefabricated concrete elements in the case of larger projects or for professional works (e.g. electricity installation).

In the first step, the materials of the WBS elements are identified and then converted to a common unit of measurement. For example, n pieces of a particular brick product with 30-centimetre thickness are built in according to the WBS, which is converted to m square metres of brick for outer walls (30 cm). The aim is to produce a spreadsheet of materials which can be easily aggregated in the second step. The aggregation is a crucial step as a material may occur in various elements. For example, expanded polystyrene (EPS) insulation can be used in various thicknesses for a project. This step allows us to identify those materials that have a major share in the total environmental impact disregarding the fact that it is dismantled in some elements.

The third step is relevant only in the case of some marginal materials, where the unit of measurement used for aggregation cannot be used for further ecological footprint calculation. For these elements, a second conversion is applied.

The analysed sample represents the major share of the Hungarian housing stock as it comprises four projects aiming to build or retrofit an SFH: (1) a new SFH with a flat roof (Szennay et al., 2021), (2) a new SFH with gable roof (Szennay et al., 2021), (3) a new SFH made of prefabricated timber elements (Szigeti et al., 2022), and (4) retrofitting an existing Hungarian cube.

Regardless of the wall materials, all analysed projects have a solid basement made of concrete, the roofs are interchangeable—apart from the new SFH with a flat roof—and provide a decent home for a family. As the main difference among the analysed projects lies in the material of the outer wall, the chapter focuses on the environmental impact of the outer wall and its insulation. This limitation also allows disregarding other factors that may be influenced by the user's discretion, for example, machinery, heating, surface covering, etc., since their price, quality, and environmental impact range widely.

Although the floor area and the energy efficiency are also similar for each project, two major differences occur. First, the new dwellings provide more living space and therefore higher living standards as the two SFHs built with conventional technologies have (1) a built-in attic (62 sqm), (2) a garage (35 sqm), and (3) a terrace (29 sqm). Second, the useful lifespan of buildings is not necessarily similar, as (1) the retrofitted Hungarian cube is a minimum of 30 years old and (2) the new SFH made of timber has a lightweight structure which can result in a shorter lifetime compared to brick structures.

As detailed data of each of the SFHs analysed have been published before, only the technical parameters of an energetic upgrade of an existing Hungarian cube are presented in this chapter (see Table 2). Based on our professional assumption, three energetic upgrade packages were created, which can be used both one by one or combined.

4 Results

Our results show that an energetic upgrade of an existing Hungarian cube has a significantly lower environmental impact than building a new one. When only the insulation is considered, the embedded CO_2 emission is only 2.8% of those of the new SFH, but this ratio lies at about 8.0% in the case of an extensive refurbishment

Table 2 Specification of the analysed projects

Package		Surface (sqm)	Thickness (cm)	Volume (m3)
Insulation	Façade: EPS insulation	105	20	21.00
	Façade: Adhesive and paint	105	0.3	0.32
	Footing: XPS insulation	24	4	0.96
	Footing: Adhesive and paint	24	0.3	0.07
	End slab: Fibre insulation (e.g. rock wool)	100	10	10.00
Replacing doors and windows	Windows (3 pieces, 200 × 150 cm)	n/r	n/r	n/r
	Windows (2 pieces, 60 × 40 cm)	n/r	n/r	n/r
	Entrance door (1 piece, 100 × 200 cm)	n/r	n/r	n/r
Roof reconstruction	Timber structure	n/r	n/r	8.946
	Particle board	n/r	n/r	0.684
	Concrete roof tiles and accessories	250	n/r	n/r

Source: own elaboration

(insulation, new roofing, windows, and doors). Considering the larger useful floor area, the specific embedded carbon lies at 76.3 versus 564.4 kg per square metre which means that a deep refurbishment with extensive technical content is viable from an environmental point of view (see Table 3). According to our expert opinion, the costs will be much lower even in the case of extension and deep renovation of an existing house.

The high embedded CO_2 figure can be explained mostly by the usage of concrete in the foundations and structural works, and to a lesser extent by the masonry works. These two work phases together account for 72.7% of the total environmental impact. Thus, our results are in line with former literature sources. It is important to emphasise, however, that as the concrete demand of an SFH with a gable roof is less than that of a flat roof, the environmental impact of the latter is significantly higher.

Although the foundation and the roofing can be the same in the case of an SFH built with non-conventional materials, using them will result in significant savings concerning the environmental impact. For example, substituting walls built of conventional bricks with prefabricated timber elements and polystyrene insulation with timber-based ones (e.g. woodwool, cellulose, etc.) will result in less than half the environmental impact (Table 4).

5 Discussion

Our results—similar to our former findings (see Szigeti et al., 2023)—show that retrofitting an existing building has a significantly lower ecological footprint than building a new one (2.32 gha vs 13.36–29.00 gha). A systematic literature review by

Table 3 Comparison of the environmental impact of an energy upgrade of a Hungarian cube and a new single-family house

Embedded carbon dioxide (tonnes)	Energy upgrade of a Hungarian cube	New SFH
Foundation	n/r	24.53
On-site concrete and reinforced concrete work	n/r	26.53
Prefabricated building construction	n/r	0.08
Masonry and other masonry work	n/r	11.29
Carpentry	2.14	2.14
Plastering and rendering	n/r	3.56
Light-structure work	n/r	2.60
Roofing	1.26	1.26
Tiling	n/r	4.15
Tinning	0.14	0.44
Insulation	2.41	7.10
Windows and doors	0.93	2.13
Total embedded CO_2 (tonnes)	**6.9**	**85.8**
Useful floor area (sqm)	90.0	152.0
Kg CO_2/sqm	76.3	564.4

Source: own elaboration

Table 4 Comparison of the environmental impact of single-family houses

Building	Embedded CO_2	Ecological footprint	References
A new SFH with a gable roof	85.79	29.00	Szennay et al. (2021)
A new SFH made of prefabricated timber elements			Szigeti et al. (2022)
Lightweight structure, walls, insulation	11.17	3.78	
Foundation, roofing, tinning, windows, and doors	28.35	9.58	
Total	39.52	13.36	
Retrofitting an existing Hungarian cube	6.87	2.32	Table 3

Source: own elaboration

Schwartz et al. (2018), however, concluded that considering the whole life cycle it is not possible to conclusively determine which one should be preferred. Another finding is that using non-conventional building materials may result in significant savings concerning environmental impact. Our example, an SFH made of prefabricated timber elements, shows that substituting only the walls with natural material will significantly decrease the total environmental impact of a newly built house. An important limitation is, however, that the expected lifetime of a timber house may differ. As the costs of such timber houses are also competitive compared to conventional ones, building timber houses can be recommended if it is done with due care.

Government programs significantly influence residents' decisions regarding home purchases and renovations. Given the high environmental impact of new construction compared to refurbishment, state incentives should prioritise energy upgrades and selectively support new construction, especially in urban revitalisation areas. A comprehensive government program could achieve both environmental goals and economic growth, as indicated by previous expert reports (Illésné Szécsi et al., 2021; Ürge-Vorsatz et al., 2010). Additionally, residents' investment decisions are influenced by macroeconomic cycles, so the availability of predictable and subsidised funds (e.g. grants and loans) can motivate energy-efficient renovations that are economically viable.

Both the changing demographic landscape of Western societies and urbanisation pose a challenge for decision-makers and urban planners. The scarcity of decent living habitats in urban zones and the need for green areas motivate moving to the agglomeration zones. This process has three major negative environmental consequences. First, commuting causes traffic congestions, bad air quality, and even accidents on the commuting roads during peak times (see, for example, Hardi et al., 2020). Second, urban sprawl means the construction of new homes on former croplands, the ecological footprint of which is extremely high (Kovács et al., 2020, 2022). Third, utilities (water and wastewater supply, roads, public lighting, etc.) and communal services (e.g. schools) cannot adapt to the increasing demand in the short term, which could result in suboptimal solutions. At the same time, rural areas can be characterised by out-migration of populations, decreasing quality of communal services and sluggish economic conditions. Such a lack of perspective will result in a vicious circle of growing and less sustainable urbanisation zones and a deprived countryside.

Our results show that non-conventional building materials and methods (e.g. adobe, straw, timber, etc.) have significantly lower environmental impact compared to conventional ones (e.g. fired bricks, concrete blocks). Efficiently using them, however, requires the education of both residents and construction firms. It must be highlighted that these materials and methods have their limitations (e.g. fire resistance, resistance to external humidity, etc.), so extensive usage needs some kind of consciousness to avoid bad experiences. A further limitation is that insurance companies do not prefer providing insurance schemes for adobe houses, so policymaking should encourage them to develop more selective risk assessment processes.

The practical implication of our results is that architects and the construction industry should consider both innovative and non-conventional building materials and technologies when designing, planning, and building new homes since literature sources provide much evidence that financial, technical, and environmental aspects are compatible. It means that the customer expectations on living standards, technical quality, durability, as well as construction and operational costs can be satisfied using other than conventional methods.

6 Summary

The chapter seeks to comprehend the environmental effects of housing in Hungary by examining typical Hungarian dwellings, including a new two-storey family house with a gable roof, a new timber family house, and an energy upgrade of a typical family dwelling from 1950 to 1990, known as the "Hungarian cube". These samples are suitable representations of single-family dwelling options in Hungary and Central and Eastern Europe.

The study findings indicate that refurbishing an existing Hungarian cube dwelling has a substantially lower environmental impact compared to constructing a new one. Specifically, when only insulation is upgraded, the embedded CO_2 emissions are a mere 2.8% of those associated with a new single-family house (SFH). However, with extensive refurbishment including insulation, roofing, windows, and doors, this ratio increases to approximately 8.0%. Despite this, the specific embedded carbon per square meter is significantly lower for deep refurbishment at 76.3 kg compared to 564.4 kg for new construction, suggesting that environmentally sound deep renovations are feasible. The primary contributors to the high embedded CO_2 are concrete usage in foundations and structural works, which collectively represent 72.7% of the total environmental impact. These findings align with prior research. Notably, SFHs with flat roofs have a higher environmental impact due to increased concrete demand. Using non-conventional materials such as prefabricated timber elements and timber-based insulation can significantly reduce environmental impact compared to conventional materials like bricks and polystyrene insulation, potentially halving the impact. Therefore, the study underscores the environmental benefits of deep renovation and alternative material use in housing construction.

The limitations of our chapter are twofold. First, our analysis does not take into consideration the operational phase of SFHs. Nevertheless, as the energy characteristics of the analysed projects are very similar, it is not expected that significant differences occur with respect to heating and cooling during the lifetime. Although the shorter useful lifetime is a frequent concern towards lightweight structure houses, empirical evidence shows that when those are handled with the necessary care, their lifetime can be similar to those built of bricks or concrete. The second limitation is that only timber SFH was considered in the analysis. Thus, further possible research can be done to analyse other non-conventional building materials and methodologies, like adobe or even straw. In this case, however, we could not acquire appropriate and well-founded data to conduct a detailed analysis of such houses.

Acknowledgements The research work was supported by the National Research, Development, and Innovation Fund of Hungary under Grant Nr. K-135546 and K-143542. Áron Szennay's contribution was supported by project no. TKP2021-NKTA-44, which was implemented with support provided by the Ministry of Innovation and Technology of Hungary from the National Research, Development and Innovation Fund, financed under the Tématerületi Kiválósági Program 2021 (TKP2021-NKTA) funding scheme.

References

Abed, M., & Nemes, R. (2019). Long-term durability of self-compacting high-performance concrete produced with waste materials. *Construction and Building Materials, 212*, 350–361. https://doi.org/10.1016/j.conbuildmat.2019.04.004

Abed, M. A., Alrefai, M., Alali, A., Nemes, R., & Yehia, S. (2022). Effect of nominal maximum aggregate size on the performance of recycled aggregate self-compacting concrete: Experimental and numerical investigation. *Materials Journal, 119*(5), 37–50. https://doi.org/10.14359/51735948

Akadiri, P. O., Chinyio, E. A., & Olomolaiye, P. O. (2012). Design of a sustainable building: A conceptual framework for implementing sustainability in the building sector. *Buildings, 2*(2), 2. https://doi.org/10.3390/buildings2020126

Alawneh, R., Mohamed Ghazali, F. E., Ali, H., & Asif, M. (2018). Assessing the contribution of water and energy efficiency in green buildings to achieve United Nations sustainable development goals in Jordan. *Building and Environment, 146*, 119–132. https://doi.org/10.1016/j.buildenv.2018.09.043

Alba-Rodríguez, M. D., Solís-Guzmán, J., & Marrero, M. (2022). Evaluation model of the economic-environmental impact on housing recovery. Application in the city of Seville, Spain. *Sustainable Cities and Society, 83*, 103940. https://doi.org/10.1016/j.scs.2022.103940

Allen, J. G., MacNaughton, P., Laurent, J. G. C., Flanigan, S. S., Eitland, E. S., & Spengler, J. D. (2015). Green buildings and health. *Current Environmental Health Reports, 2*(3), 250–258. https://doi.org/10.1007/s40572-015-0063-y

AlSanad, S. (2015). Awareness, drivers, actions, and barriers of sustainable construction in Kuwait. *Procedia Engineering, 118*, 969–983. https://doi.org/10.1016/j.proeng.2015.08.538

Ashour, T., Georg, H., & Wu, W. (2011). Performance of straw bale wall: A case of study. *Energy and Buildings, 43*(8), 1960–1967. https://doi.org/10.1016/j.enbuild.2011.04.001

Beleznay, É. (2021). *Szakértői tanulmány az építési ágazat karbon lábnyom csökkentésére* (NFFT Műhelytanulmány). Nemzeti Fenntartható Fejlődési Tanács. https://www.nfft.hu/documents/1238941/0/NFFT_Epitoipar_20210316+%281%29.pdf/eb8d7060-6829-48a4-0926-37f1cc3400da?t=1621501854096

Berrone, P., Fosfuri, A., Gelabert, L., & Gomez-Mejia, L. R. (2013). Necessity as the mother of 'green' inventions: Institutional pressures and environmental innovations. *Strategic Management Journal, 34*(8), 891–909.

Buyle, M., Braet, J., & Audenaert, A. (2013). Life cycle assessment in the construction sector: A review. *Renewable and Sustainable Energy Reviews, 26*, 379–388. https://doi.org/10.1016/j.rser.2013.05.001

Cao, X., Dai, X., & Liu, J. (2016). Building energy-consumption status worldwide and the state-of-the-art technologies for zero-energy buildings during the past decade. *Energy and Buildings, 128*, 198–213. https://doi.org/10.1016/j.enbuild.2016.06.089

Cao, S., Ma, Z., Liu, Z., Guo, J., & Yuan, W. (2023). Achieving harmony between the economy and the environment. *Habitat International, 131*, 102733. https://doi.org/10.1016/j.habitatint.2022.102733

Chau, C. K., Hui, W. K., Ng, W. Y., & Powell, G. (2012). Assessment of CO2 emissions reduction in high-rise concrete office buildings using different material use options. *Resources, Conservation and Recycling, 61*, 22–34. https://doi.org/10.1016/j.resconrec.2012.01.001

European Commission. (n.d.). *Long-term renovation strategies*. Retrieved March 21, 2024, from https://energy.ec.europa.eu/topics/energy-efficiency/energy-efficient-buildings/long-term-renovation-strategies_en

European Commission. Directorate General for Economic and Financial Affairs. (2023). *2023 country report: Hungary*. Publications Office. https://data.europa.eu/doi/10.2765/354358

European Energy Agency. (2023). *The benefits to biodiversity of a strong circular economy*. https://www.eea.europa.eu/publications/the-benefits-to-biodiversity/the-benefits-to-biodiversity-of/download.pdf.static

Eurostat. (n.d.). *Housing in Europe—House or flat—Owning or renting*. Housing in Europe. Retrieved January 10, 2024, from https://ec.europa.eu/eurostat/cache/digpub/housing/bloc-1a.html

Foster, J., Lowe, A., & Winkelman, S. (2011). *The value of green infrastructure for urban climate adaptation*. The Center for Clean Air Policy. https://savetherain.us/wp-content/uploads/2011/10/Green_Infrastructure_Urban_Climate_Adaptation.pdf

González, A. D. (2014). Energy and carbon embodied in straw and clay wall blocks produced locally in the Andean Patagonia. *Energy and Buildings, 70*, 15–22. https://doi.org/10.1016/j.enbuild.2013.11.003

Gyurkó, Z., Szijártó, A., & Nemes, R. (2019). Cellular concrete waste as an economical alternative to traditional supplementary cementitious materials. *Journal of Thermal Analysis and Calorimetry, 138*(2), 947–961. https://doi.org/10.1007/s10973-019-08303-8

Hardi, T., Repaská, G., Veselovský, J., & Vilinová, K. (2020). Environmental consequences of the urban sprawl in the suburban zone of Nitra: An analysis based on landcover data. *Geographica Pannonica, 24*(3), 205–220. https://doi.org/10.5937/gp24-25543

Hauashdh, A., Nagapan, S., Jailani, J., & Gamil, Y. (2024). An integrated framework for sustainable and efficient building maintenance operations aligning with climate change, SDGs, and emerging technology. *Results in Engineering, 21*, 101822. https://doi.org/10.1016/j.rineng.2024.101822

HCSO. (n.d.). *Census 2022*. Census 2022. Retrieved January 11, 2024, from https://nepszamlalas2022.ksh.hu/en/index

He, X., Gao, W., Guan, D., & Zhou, L. (2023). Impacts of urban shrinkage on the built environment and its environmental sustainability: An analytical review. *Environmental Research Letters, 18*(10), 103004. https://doi.org/10.1088/1748-9326/acf726

Huang, L., Krigsvoll, G., Johansen, F., Liu, Y., & Zhang, X. (2018). Carbon emission of global construction sector. *Renewable and Sustainable Energy Reviews, 81*, 1906–1916. https://doi.org/10.1016/j.rser.2017.06.001

Illésné Szécsi, I., Koritár, Z., Pálffy, A., & Sáfián, F. (2021). *Hazai felújítási hullám. A magyarországi lakóépület-állomány energiahatékonysági korszerűsítésében rejlő lehetőségek, egyes támogatási eszközök széles körű hatásainak vizsgálata*. Magyar Energiahatékonysági Intézet. https://mehi.hu/wp-content/uploads/2021/03/mehi_hazai_felujitasi_hullam_tanulmany_2021_v3_0.pdf

Interreg Europe. (n.d.). *Sustainability in construction | Interreg Europe—Sharing solutions for better policy*. Retrieved March 21, 2024, from https://www.interregeurope.eu/event/promoting-sustainable-materials-and-circularity-in-construction

Jaradat, H., Alshboul, O. A. M., Obeidat, I. M., & Zoubi, M. K. (2024). Green building, carbon emission, and environmental sustainability of construction industry in Jordan: Awareness, actions and barriers. *Ain Shams Engineering Journal, 15*(2), 102441. https://doi.org/10.1016/j.asej.2023.102441

Jia, H. J., & Wang, H. (2012). Green building and sustainable development. *Advanced Materials Research, 575*, 55–58. https://doi.org/10.4028/www.scientific.net/AMR.575.55

Kisku, N., Joshi, H., Ansari, M., Panda, S. K., Nayak, S., & Dutta, S. C. (2017). A critical review and assessment for usage of recycled aggregate as sustainable construction material. *Construction and Building Materials, 131*, 721–740. https://doi.org/10.1016/j.conbuildmat.2016.11.029

Kocsis, T. (2014). Is the Netherlands sustainable as a global-scale inner-city? Intenscoping spatial sustainability. *Ecological Economics, 101*, 103–114. https://doi.org/10.1016/j.ecolecon.2014.03.002

Kovács, Z., Harangozó, G., Szigeti, C., Koppány, K., Kondor, A. C., & Szabó, B. (2020). Measuring the impacts of suburbanization with ecological footprint calculations. *Cities, 101*, 102715. https://doi.org/10.1016/j.cities.2020.102715

Kovács, Z., Farkas, J. Z., Szigeti, C., & Harangozó, G. (2022). Assessing the sustainability of urbanization at the sub-national level: The ecological footprint and biocapacity accounts of the Budapest metropolitan region, Hungary. *Sustainable Cities and Society, 84*, 104022. https://doi.org/10.1016/j.scs.2022.104022

Li, H., Luo, Z., Xu, X., Cang, Y., & Yang, L. (2021). Assessing the embodied carbon reduction potential of straw bale rural houses by hybrid life cycle assessment: A four-case study. *Journal of Cleaner Production, 303*, 127002. https://doi.org/10.1016/j.jclepro.2021.127002

Lupton, K., & Samy, C. (2022). Restoring the balance between humanity and nature through tourism entrepreneurship: A conceptual framework. *Journal of Tourism Futures, 8*(3), 367–374. https://doi.org/10.1108/JTF-01-2022-0035

Marrero, M., Rivero-Camacho, C., Martínez-Rocamora, A., Alba-Rodríguez, D., & Lucas-Ruiz, V. (2024). Holistic assessment of the economic, environmental, and social impact of building construction. Application to housing construction in Andalusia. *Journal of Cleaner Production, 434*, 140170. https://doi.org/10.1016/j.jclepro.2023.140170

Matolcsy, K. (2020). *Kockaházak felújítási kérdései*. https://epitesimegoldasok.hu/kockahazak-felujitasi-kerdesei.html

Mengistu, G. M., & Nemes, R. (2024). Recycling 3D printed concrete waste for normal strength concrete production. *Applied Sciences, 14*(3), 3. https://doi.org/10.3390/app14031142

Mirasgedis, S., Cabeza, L. F., & Vérez, D. (2024). Contribution of buildings climate change mitigation options to sustainable development. *Sustainable Cities and Society, 106*, 105355. https://doi.org/10.1016/j.scs.2024.105355

OECD. (2023). *Towards a national circular economy strategy for Hungary*. https://www.oecd.org/env/waste/Highlights-Towards-a-National-Circular-Economy-Strategy-for-Hungary_EN.pdf

Omer, M. A. B., & Noguchi, T. (2020). A conceptual framework for understanding the contribution of building materials in the achievement of sustainable development goals (SDGs). *Sustainable Cities and Society, 52*, 101869. https://doi.org/10.1016/j.scs.2019.101869

Pitt, M., Tucker, M., Riley, M., & Longden, J. (2009). Towards sustainable construction: Promotion and best practices. *Construction Innovation, 9*(2), 201–224. https://doi.org/10.1108/14714170910950830

Sandanayake, M. S. (2022). Environmental impacts of construction in building industry—A review of knowledge advances, gaps and future directions. *Knowledge, 2*(1), 1. https://doi.org/10.3390/knowledge2010008

Schwartz, Y., Raslan, R., & Mumovic, D. (2018). The life cycle carbon footprint of refurbished and new buildings—A systematic review of case studies. *Renewable and Sustainable Energy Reviews, 81*, 231–241. https://doi.org/10.1016/j.rser.2017.07.061

Sharifi, A., Allam, Z., Bibri, S. E., & Khavarian-Garmsir, A. R. (2024). Smart cities and sustainable development goals (SDGs): A systematic literature review of co-benefits and trade-offs. *Cities, 146*, 104659. https://doi.org/10.1016/j.cities.2023.104659

Sodagar, B., Rai, D., Jones, B., Wihan, J., & Fieldson, R. (2011). The carbon-reduction potential of straw-bale housing. *Building Research & Information, 39*(1), 51–65. https://doi.org/10.1080/09613218.2010.528187

Solís-Guzmán, J., Martínez-Rocamora, A., & Marrero, M. (2014). Methodology for determining the carbon footprint of the construction of residential buildings. In S. S. Muthu (Ed.), *Assessment of carbon footprint in different industrial sectors, volume 1* (pp. 49–83). Springer. https://doi.org/10.1007/978-981-4560-41-2_3

Stanitsas, M., Kirytopoulos, K., & Leopoulos, V. (2021). Integrating sustainability indicators into project management: The case of construction industry. *Journal of Cleaner Production, 279*, 123774. https://doi.org/10.1016/j.jclepro.2020.123774

Szalay, Z., & Gröller, G. (2021). Megéri-e egy Kádár-kocka felújítása környezeti szempontból? *Építészfórum*. https://epiteszforum.hu/megeri-e-egy-kadar-kocka-felujitasa-kornyezeti-szempontbol

Szalay, Z., Szagri, D., Bihari, Á., Nagy, B., Kiss, B., Horváth, M., & Medgyasszay, P. (2022). Development of a life cycle net zero carbon compact house concept. *Energy Reports, 8*, 12987–13013. https://doi.org/10.1016/j.egyr.2022.09.197

Szécsi, I. (2021). *A Kádár-kockákban rejlő energiahatékonysági felújítási potenciál*. https://www.epiteszforum.hu. https://epiteszforum.hu/a-kadar-kockakban-rejlo-energiahatekonysagi-felujitasi-potencial

Szennay, Á., Major, Z., & Beke, J. (2021). *Ecological footprint satellite calculators to determine the environmental impact of material usage of SMEs*. In 12th IEEE International conference on cognitive infocommunications (CogInfoCom 2021): Proceedings (pp. 677–680). https://www.researchgate.net/publication/365841797_Ecological_footprint_satellite_calculators_to_determine_the_environmental_impact_of_material_usage_of_SMEs

Szép, J., Major, Z., Szigeti, C., & Lubloy, É. (2023). The impact of cement aggregates on the fire resistance properties of concrete and its ecological footprint. *Chemical Engineering Transactions, 107*, 337–342. https://doi.org/10.3303/CET23107057

Szigeti, C., Szennay, Á., & Major, Z. (2022). Épületek ökológiai lábnyoma: Fából vagy betonból építsünk? In K. Szegedi (Ed.), *Szemelvények a BGE kutatásaiból* (pp. 212–219). Budapesti Gazdasági Egyetem. https://doi.org/10.29180/978-615-6342-49-2_25

Szigeti, C., Major, Z., Szabó, D. R., & Szennay, Á. (2023). The ecological footprint of construction materials—A standardized approach from Hungary. *Resources, 12*(1), 15. https://doi.org/10.3390/resources12010015

Talapin, D. V., & Braun, P. V. (2020). EC announces actions to increase security and sustainability of Europe's raw materials supply. *MRS Bulletin, 45*(10), 796. https://doi.org/10.1557/mrs.2020.277

Tamáska, M. (2021). A kockaház. Kutatási előzmények és perspektívák. In M. Tamáska (Ed.), *Kockaház: A 20. Század vidéki háztípusa*. Martin Opitz Kiadó.

United Nations. (2015). *Transforming our world: The 2030 Agenda for Sustainable Development*. https://www.un.org/ga/search/view_doc.asp?symbol=A/RES/70/1&Lang=E

United Nations Environment Programme, Y. C. for E. + A. (2023). *Building materials and the climate: Constructing a new future*. https://wedocs.unep.org/20.500.11822/43293

Ürge-Vorsatz, D., & Novikova, A. (2008). Potentials and costs of carbon dioxide mitigation in the world's buildings. *Energy Policy, 36*(2), 642–661. https://doi.org/10.1016/j.enpol.2007.10.009

Ürge-Vorsatz, D., Arena, D., Tirado Herrero, S., & Butcher, A. (2010). *Egy nagyszabású, energia-megtakarítást célzó, komplex épületfelújítási program hatása a foglalkoztatásra Magyarországon*. Center for Climate Change and Sustainable Energy Policy (3CSEP) Central European University. https://envsci.ceu.edu/sites/envsci.ceu.hu/files/attachment/project/558/magyarfullreport.pdf

Vasárus, G. L., & Lennert, J. (2022). Suburbanization within city limits in Hungary—A challenge for environmental and social sustainability. *Sustainability*, 14(14), Article 14. https://doi.org/10.3390/su14148855.

World Economic Forum. (2016). *Shaping the future of construction. A breakthrough in mindset and technology*. https://www3.weforum.org/docs/WEF_Shaping_the_Future_of_Construction_full_report__.pdf

World Green Building Week. (2023). *The sustainable transition depends on scaling positive change now! WorldGBC announces #BuildingTheTransition for World Green Building Week*. World Green Building Council. https://worldgbc.org/article/buildingthetransition-world-green-building-week-2023/

Wu, Z., Jiang, M., Cai, Y., Wang, H., & Li, S. (2019). What hinders the development of green building? An investigation of China. *International Journal of Environmental Research and Public Health, 16*(17), 17. https://doi.org/10.3390/ijerph16173140

Yang, H., Gong, Z., Chen, C., & Zhao, T. (2020). Ecological footprint of residential building in Xining. *IOP Conference Series: Earth and Environmental Science, 495*(1), 012052. https://doi.org/10.1088/1755-1315/495/1/012052

Zalatnay, L. (2016). *Szalmabála építészet és környezettudatos megoldásai felhasználási lehetőségei a közösségi energiagazdálkodás terén Tanulmány*. Magyar Természetvédők Szövetsége. https://mtvsz.hu/uploads/files/kozossegi_energia_szalmaepiteszet_tanulmany_mtvsz.pdf

Zhang, X., Shen, L., & Zhang, L. (2013). Life cycle assessment of the air emissions during building construction process: A case study in Hong Kong. *Renewable and Sustainable Energy Reviews, 17*, 160–169. https://doi.org/10.1016/j.rser.2012.09.024

Aron Szennay (PhD) is a senior research fellow at the Budapest LAB Office for Entrepreneurship Development, Budapest Business University, Budapest, Hungary. His research, publication, and teaching activities focus on the responsibility of business sectors, sustainability, and entrepreneurship. Among the main research topics are environmental impact calculation of businesses. His PhD thesis was based on a multiannual research program aimed at developing a standardised online ecological footprint calculator. The current, main research area is concerned with entrepreneurship, among others family firms. He is the author of several publications on this subject.

Zoltán Major (PhD) is an assistant professor at Széchenyi István University in Győr, Hungary. His research, publication, and teaching activities focus on civil engineering sciences. His main research interests include the planning and building of railway infrastructure. His main research area is the designing of embedded rail structures.

Róbert Horváth (MSc) is the managing director of the engineering firm H-Planer Mérnökiroda Kft and a PhD student at the Széchenyi István University in Győr. His research, publication, and teaching activities are in the field of civil engineering. His main field of research is the design, construction, and maintenance of railway infrastructure. His doctoral thesis focuses on the rail-wheel contact.

Herta Czédli (PhD) is an associate professor at the University of Debrecen in Debrecen, Hungary. Her research, publication, and teaching activities focus on environmental and urban systems engineering. Her main research area is the analysis of anthropogenic effects on the environment. Her main research areas are the analysis of urban ecological problems and the development of new appropriate ecofriendly technological solutions that provide answers to specific urban environmental challenges.

Cecilia Szigeti (PhD, Hab.) is the head of the Institute for Sustainability Studies at Budapest Metropolitan University in Budapest, Hungary. Her research, publication, and teaching activities focus on the interrelationship between economy, society, and the natural environment. Her main research interests include environmental economics and management and its interfaces. Her main research area is ecological footprinting.

Index

B
Brazil, 2, 24–34

C
Carbon neutral, 14, 23, 109, 113
City energy, 38, 52–55, 58, 61
Computers, 3, 65–102

D
Decision-making, 3, 9, 13, 65–68, 70–75, 77–78, 84, 86–92, 98, 99, 101, 102

E
Ecological footprint (EF), 3, 77, 78, 87–89, 94, 96, 114–117, 119, 120
Electric vehicles, 47, 49
Energy, 1, 5, 23, 37, 65, 105
Energy footprints, 1–17, 23–34, 37–63, 65–92, 96, 100–102
Environmental challenges, 67, 81, 82, 85–87, 92–101
Environmental impacts, 3, 4, 14, 15, 34, 65–102, 105, 106, 108–121

F
Family houses, 3, 105–121

H
Housing, 2, 3, 37–46, 49–52, 54–63, 105–121

M
Measurement models, 11
Mitigation, 11, 29, 69, 106, 113

N
National Hydrogen Program, 2, 28, 30

R
Renewable hydrogen, 2, 23–34
Review, 1, 2, 5–17, 24, 25, 46, 61, 96, 106, 115, 119
Ride-hailing, 3, 62

S
Self-driving cars, 3
Sustainability, 1–4, 9, 10, 14, 17, 27, 31, 33, 66, 67, 71, 72, 74–78, 80–83, 85–88, 90, 92–94, 96–101, 106, 109, 111–113
Sustainable Development Goals (SDGs), 4, 14, 24, 106

S. S. Muthu (ed.), *Energy Footprint and Sustainability*, Environmental Footprints and Eco-design of Products and Processes,
https://doi.org/10.1007/978-3-031-63057-6

T
Telework, 3, 60, 62
Transportation, 2, 7, 15, 24, 25, 31, 34, 39–42, 44–46, 49, 51, 53, 54, 62, 71, 94, 96, 97

W
Work from home (WFH), 3, 60–61

Z
Zoning, 2

SPRINGER NATURE

GPSR Compliance

The European Union's (EU) General Product Safety Regulation (GPSR) is a set of rules that requires consumer products to be safe and our obligations to ensure this.

If you have any concerns about our products, you can contact us on ProductSafety@springernature.com

In case Publisher is established outside the EU, the EU authorized representative is:

Springer Nature Customer Service Center GmbH
Europaplatz 3
69115 Heidelberg, Germany

The manufacturer's authorised representative in the EU is Springer Nature Customer Service Centre GmbH, Europaplatz 3, 69115 Heidelberg, Germany. If you have any concerns regarding our products, please contact ProductSafety@springernature.com

Printed and bound by CPI Group (UK) Ltd, Croydon, CR0 4YY
30/09/2026
02225604-0006